Voyages in Search of the North-West Passage

Richard Hakluyt

(Editor: Henry Morley)

Alpha Editions

This edition published in 2024

ISBN : 9789364736084

Design and Setting By
Alpha Editions
www.alphaedis.com
Email - info@alphaedis.com

As per information held with us this book is in Public Domain.
This book is a reproduction of an important historical work. Alpha Editions uses the best technology to reproduce historical work in the same manner it was first published to preserve its original nature. Any marks or number seen are left intentionally to preserve its true form.

Contents

INTRODUCTION.	- 1 -
A DISCOURSE WRITTEN BY SIR HUMPHREY GILBERT, KNIGHT.	- 14 -
CERTAIN OTHER REASONS OR ARGUMENTS TO PROVE A PASSAGE BY THE NORTH-WEST.	- 36 -
THE FIRST VOYAGE OF MASTER MARTIN FROBISHER	- 45 -
THE SECOND VOYAGE OF MASTER MARTIN FROBISHER,	- 50 -
THE THIRD AND LAST VOYAGE INTO META INCOGNITA,	- 64 -
THE REPORT OF THOMAS WIARS,	- 73 -
THE FIRST VOYAGE OF MASTER JOHN DAVIS,	- 74 -
THE SECOND VOYAGE ATTEMPTED BY MASTER JOHN DAVIS,	- 83 -
THE THIRD VOYAGE NORTH-WESTWARD, MADE BY JOHN DAVIS,	- 97 -

INTRODUCTION.

Thirty-five years ago I made a voyage to the Arctic Seas in what Chaucer calls

> A little bote
> No bigger than a mannë's thought;

it was a Phantom Ship that made some voyages to different parts of the world which were recorded in early numbers of Charles Dickens's "Household Words." As preface to Richard Hakluyt's records of the first endeavour of our bold Elizabethan mariners to find North-West Passage to the East, let me repeat here that old voyage of mine from No. 55 of "Household Words," dated the 12th of April, 1851: The *Phantom* is fitted out for Arctic exploration, with instructions to find her way, by the north-west, to Behring Straits, and take the South Pole on her passage home. Just now we steer due north, and yonder is the coast of Norway. From that coast parted Hugh Willoughby, three hundred years ago; the first of our countrymen who wrought an ice-bound highway to Cathay. Two years afterwards his ships were found, in the haven of Arzina, in Lapland, by some Russian fishermen; near and about them Willoughby and his companions—seventy dead men. The ships were freighted with their frozen crews, and sailed for England; but, "being unstaunch, as it is supposed, by their two years' wintering in Lapland, sunk, by the way, with their dead, and them also that brought them."

Ice floats about us now, and here is a whale blowing; a whale, too, very near Spitzbergen. When first Spitzbergen was discovered, in the good old times, there were whales here in abundance; then a hundred Dutch ships, in a crowd, might go to work, and boats might jostle with each other, and the only thing deficient would be stowage room for all the produce of the fishery. Now one ship may have the whole field to itself, and travel home with an imperfect cargo. It was fine fun in the good old times; there was no need to cruise. Coppers and boilers were fitted on the island, and little colonies about them, in the fishing season, had nothing to do but tow the whales in, with a boat, as fast as they were wanted by the copper. No wonder that so enviable a Tom Tidler's ground was claimed by all who had a love for gold and silver. The English called it theirs, for they first fished; the Dutch said, nay, but the island was of their discovery; Danes, Hamburghers, Bisayans, Spaniards, and French put in their claims; and at length it was agreed to make partitions. The numerous bays and harbours which indent the coast were divided among the rival nations; and, to this day, many of them bear, accordingly, such names as English Bay, Danes Bay, and so forth. One bay there is, with graves in it, named Sorrow. For it seemed to the fishers most desirable, if possible, to plant upon this island permanent establishments, and condemned convicts were offered, by the Russians, life and pardon, if they would winter in Spitzbergen. They agreed; but, when they saw the icy mountains and the stormy sea, repented, and went back, to meet a death exempt from torture. The Dutch tempted free men, by high rewards, to try the dangerous experiment. One of their victims left a journal, which describes his suffering and that of his companions. Their mouths, he says, became so sore that, if they had food, they could not eat; their limbs were swollen and disabled with excruciating pain; they died of scurvy. Those who died first were coffined by their dying friends; a row of coffins was found, in the spring, each with a man in it; two men uncoffined, side by side, were dead upon the floor. The journal told how once the traces of a bear excited their hope of fresh meat and amended health; how, with a lantern, two or three had limped upon the track, until the light became extinguished, and they came back in despair to die. We might speak, also, of eight English sailors, left, by accident, upon Spitzbergen, who lived to return and tell their winter's tale; but a long journey is before us and we must not linger on the way. As for our whalers, it need scarcely be related that the multitude of whales diminished as the slaughtering went on, until it was no longer possible to keep the coppers full. The whales had to be searched for by the vessels, and thereafter it was not worth while to take the blubber to Spitzbergen to be boiled; and the different nations, having carried home their coppers, left the apparatus of those fishing stations to decay.

Take heed. There is a noise like thunder, and a mountain snaps in two. The upper half comes, crashing, grinding, down into the sea, and loosened streams of water follow it. The sea is displaced before the mighty heap; it boils and scatters up a cloud of spray; it rushes back, and violently beats upon the shore. The mountain rises from its bath, sways to and fro, while water pours along its mighty sides; now it is tolerably quiet, letting crackers off as air escapes out of its cavities. That is an iceberg, and in that way are all icebergs formed. Mountains of ice formed by rain and snow—grand Arctic glaciers, undermined by the sea or by accumulation over-balanced—topple down upon the slightest provocation (moved by a shout, perhaps), and where they float, as this black-looking fellow does, they need deep water. This berg in height is about ninety feet, and a due balance requires that a mass nine times as large as the part visible should be submerged. Icebergs are seen about us now which rise two hundred feet above the water's level.

There are above head plenty of aquatic birds; ashore, or on the ice, are bears, foxes, reindeer; and in the sea there are innumerable animals. We shall not see so much life near the North Pole, that is certain. It would be worth while to go ashore upon an islet there, near Vogel Sang, to pay a visit to the eider-ducks. Their nests are so abundant that one cannot avoid treading on them. When the duck is driven by a hungry fox to leave her eggs, she covers them with down, in order that they may not cool during her absence, and, moreover, glues the down into a case with a secretion supplied to her by Nature for that purpose. The deserted eggs are safe, for that secretion has an odour very disagreeable to the intruder's nose.

We still sail northward, among sheets of ice, whose boundaries are not beyond our vision from the masthead—these are "floes;" between them we find easy way, it is fair "sailing ice." In the clear sky to the north a streak of lucid white light is the reflection from an icy surface; that is, "ice-blink," in the language of these seas. The glare from snow is yellow, while open water gives a dark reflection.

Northward still; but now we are in fog the ice is troublesome; a gale is rising. Now, if our ship had timbers they would crack, and if she had a bell it would be tolling; if we were shouting to each other we should not hear, the sea is in a fury. With wild force its breakers dash against a heaped-up wall of broken ice, that grinds and strains and battles fiercely with the water. This is "the pack," the edge of a great ice-field broken by the swell. It is a perilous and an exciting thing to push through pack ice in a gale.

Now there is ice as far as eye can see, that is "an ice-field." Masses are forced up like colossal tombstones on all sides; our sailors call them "hummocks;" here and there the broken ice displays large "holes of water." Shall we go on? Upon this field, in 1827, Parry adventured with his men to reach the North Pole, if that should be possible. With sledges and portable boats they laboured on through snow and over hummocks, launching their boats over the larger holes of water. With stout hearts, undaunted by toil or danger, they went boldly on, though by degrees it became clear to the leaders of the expedition that they were almost like mice upon a treadmill cage, making a great expenditure of leg for little gain. The ice was floating to the south with them, as they were walking to the north; still they went on. Sleeping by day to avoid the glare, and to get greater warmth during the time of rest, and travelling by night—watch-makers' days and nights, for it was all one polar day—the men soon were unable to distinguish noon from midnight. The great event of one day on this dreary waste was the discovery of two flies upon an ice hummock; these, says Parry, became at once a topic of ridiculous importance. Presently, after twenty-three miles' walking, they had only gone one mile forward, the ice having industriously floated twenty-two miles in the opposite direction; and then, after walking forward eleven miles, they found themselves to be three miles behind the place from which they started. The party accordingly returned, not having reached the Pole, not having reached the eighty-third parallel, for the attainment of which there was a reward of a thousand pounds held out by government. They reached the parallel of eighty-two degrees forty-five minutes, which was the most northerly point trodden by the foot of man.

From that point they returned. In those high latitudes they met with a phenomenon, common in alpine regions, as well as at the Pole, red snow; the red colour being caused by the abundance of a minute plant, of low development, the last dweller on the borders of the vegetable kingdom. More interesting to the sailors was a fat she bear which they killed and devoured with a zeal to be repented of; for on reaching navigable sea, and pushing in their boats to Table Island, where some stones were left, they found that the bears had eaten all their bread, whereon the men agreed that "Bruin was now square with them." An islet next to Table Island—they are both mere rocks—is the most northern land discovered. Therefore, Parry applied to it the name of lieutenant—afterwards Sir James—Ross. This compliment Sir James Ross acknowledged in the most emphatic manner, by discovering on his part, at the other Pole, the most southern land yet seen, and giving to it the name of Parry: "Parry Mountains."

It very probably would not be difficult, under such circumstances as Sir W. Parry has since recommended, to reach the North Pole along this route. Then (especially if it be true, as many believe, that there is a region of open sea about the Pole itself) we might find it as easy to reach Behring Straits by travelling in a straight line over the North Pole, as by threading the straits and bays north of America.

We turn our course until we have in sight a portion of the ice-barred eastern coast of Greenland, Shannon Island. Somewhere about this spot in the seventy-fifth parallel is the most northern part of that coast known to us. Colonel—then Captain—Sabine in the *Griper* was landed there to make magnetic, and other observations; for the same purpose he had previously visited Sierra Leone. That is where we differ from our forefathers. They commissioned hardy seamen to encounter peril for the search of gold ore, or for a near road to Cathay; but our peril is encountered for the gain of knowledge, for the highest kind of service that can now be rendered to the human race.

Before we leave the Northern Sea, we must not omit to mention the voyage by Spitzbergen northward, in 1818, of Captain Buchan in the *Dorothea*, accompanied by Lieutenant Franklin, in the *Trent*. It was Sir John Franklin's first voyage to the Arctic regions. This trip forms the subject of a delightful book by Captain Beechey.

On our way to the south point of Greenland we pass near Cape North, a point of Iceland. Iceland, we know, is the centre of a volcanic region, whereof Norway and Greenland are at opposite points of the circumference. In connection with this district there is a remarkable fact; that by the agency of subterranean forces, a large portion of Norway and Sweden is being slowly upheaved. While Greenland, on the west coast, as gradually sinks into the sea, Norway rises at the rate of about four feet in a century. In Greenland, the sinking is so well known that the natives never build close to the water's edge, and the Moravian missionaries more than once have had to move farther inland the poles on which their boats are rested.

Our Phantom Ship stands fairly now along the western coast of Greenland into Davis Straits. We observe that upon this western coast there is, by a great deal, less ice than on the eastern. That is a rule generally. Not only the configuration of the straits and bays, but also the earth's rotation from west to east, causes the currents here to set towards the west, and wash the western coasts, while they act very little on the eastern. We steer across Davis Strait, among "an infinite number of great countreys and islands of yce;" there, near the entrance, we find Hudson Strait, which does not now concern us. Islands probably separate this well-known channel from Frobisher Strait to the north of it, yet unexplored. Here let us recall to mind the fleet of fifteen sail, under Sir Martin Frobisher, in 1578, tossing about and parting company among the ice. Let us remember how the crew of the *Anne Frances*, in that expedition, built a pinnace when their vessel struck upon a rock, stock, although they wanted main timber and nails. How they made a mimic forge, and "for the easier making of nails, were forced to break their tongs, gridiron, and fire-shovel, in pieces." How Master Captain Best, in this frail bark, with its imperfect timbers held together by the metamorphosed gridiron and fire-shovel, continued in his duty, and did depart up the straights as before was pretended." How a terrific storm arose, and the fleet parted and the intrepid captain was towed "in his small pinnesse, at the stern of the *Michael*, thorow the raging seas; for the bark was not able to receive, or relieve half his company." The "tongs, gridyron, and fire-shovell," performed their work only for as many minutes as were absolutely necessary, for the pinnesse came no sooner aboard the ship, and the men entred, but she presently shivered and fell in pieces, and sunke at the ship's stern with all the poor men's furniture."

Now, too, as we sail up the strait, explored a few years after these events by Master John Davis, how proudly we remember him as a right worthy forerunner of those countrymen of his and ours who since have sailed over his track. Nor ought we to pass on without calling to mind the melancholy fate, in 1606, of Master John Knight, driven, in the *Hopewell*, among huge masses of ice with a tremendous surf, his rudder knocked away, his ship half full of water, at the entrance to these straits. Hoping to find a harbour, he set forth to explore a large island, and landed, leaving two men to watch the boat, while he, with three men and the mate, set forth and disappeared over a hill. For thirteen hours the watchers kept their post; one had his trumpet with him, for he was a trumpeter, the other had a gun. They trumpeted often and loudly; they fired, but no answer came. They watched ashore all night for the return of their captain and his party, "but they came not at all."

The season is advanced. As we sail on, the sea steams like a line-kiln, "frost-smoke" covers it. The water, cooled less rapidly, is warmer now than the surrounding air, and yields this vapour in consequence. By the time our vessel has reached Baffin's Bay, still coasting along Greenland, in addition to old floes and bergs, the water is beset with "pancake ice." That is the young ice when it first begins to cake upon the surface. Innocent enough it seems, but it is sadly clogging to the ships. It sticks about their sides like treacle on a fly's wing; collecting unequally, it destroys all equilibrium, and impedes the efforts of the steersman. Rocks split on the Greenland coast with loud explosions, and more icebergs fall. Icebergs we soon shall take our leave of; they are only found where there is a coast on which glaciers can form; they are good for nothing but to yield fresh water to the vessels; it will be all field, pack, and saltwater ice presently.

Now we are in Baffin's Bay, explored in the voyages of Bylot and Baffin, 1615-16. When, in 1817, a great movement in the Greenland ice caused many to believe that the northern passages would be found comparatively clear; and when, in consequence of this impression, Sir John Barrow succeeded in setting afoot that course of modern Arctic exploration which has been continued to the present day, Sir John Ross was the first man sent to find the North-West Passage. Buchan and Parry were commissioned at the same the to attempt the North Sea route. Sir John Ross did little more on that occasion than effect a survey of Baffin's Bay, and prove the accuracy of the ancient pilot. In the extreme north of the bay there is an inlet or a channel, called by Baffin Smith's Sound; this Sir John saw, but did not enter. It never yet has been explored. It may be an inlet only; but it is also very possible that by this channel ships might get into the Polar Sea and sail by the north shore of Greenland to Spitzbergen. Turning that corner, and descending along the western coast of Baffin's Bay, there is another inlet called Jones' Sound by Baffin, also unexplored. These two inlets, with their very British titles, Smith and Jones, are of exceeding interest. Jones' Sound may lead by a back way to Melville Island. South of Jones' Sound there is a wide break in the shore, a great sound, named by Baffin, Lancaster's, which Sir John Ross, in that first expedition, failed also to explore. Like our transatlantic friends at the South Pole, he laid down a range of clouds as mountains, and considered the way impervious; so he came home. Parry went out next year, as a lieutenant, in command of his first and most successful expedition. He sailed up Lancaster Sound, which was in that year (1819) unusually clear of ice; and he is the discoverer whose track we now follow in our Phantom Ship. The whole ground being new, he had to name the points of country right and left of him. The way was broad and open, due west, a most prosperous beginning for a North-West Passage. If this continued, he would soon reach Behring Strait. A broad channel to the right, directed, that is to say, southward, he entered on the Prince of Wales's birthday, and so called it the "Prince Regent's Inlet." After exploring this for some miles, he turned back to resume his western course, for still there was a broad strait leading westward. This second part of Lancaster Sound he called after the Secretary of the Admiralty who had so indefatigably laboured to promote the expeditions, Barrow's Strait. Then he came to a channel, turning to the right or northward, and he named that Wellington Channel. Then he had on his right hand ice, islands large and small, and intervening channels; on the left, ice, and a cape visible, Cape Walker. At an island, named after the First Lord of the Admiralty Melville Island, the great frozen wilderness barred farther progress. There he wintered. On the coast of Melville Island they had passed the latitude of one hundred and ten degrees, and the men had become entitled to a royal bounty of five thousand pounds. This group of islands Parry called North Georgian, but they are usually called by his own name, Parry Islands. This was the first European winter party in the Arctic circle. Its details are familiar enough. How the men cut in three days, through ice seven inches thick, a canal two miles and a half long, and so brought the ships into safe harbour. How the genius of Parry equalled the occasion; how there was established a theatre and a *North Georgian Gazette*, to cheer the tediousness of a night which continued for two thousand hours. The dreary, dazzling waste in which there was that little patch of life, the stars, the fog, the moonlight, the glittering wonder of the northern lights, in which, as Greenlanders believe, souls of the wicked dance tormented, are familiar to us. The she-bear stays at home; but the he-bear hungers, and looks in vain for a stray seal or walrus—woe to the unarmed man who meets him in his hungry mood! Wolves are abroad, and pretty white arctic foxes. The reindeer have sought other pasture-ground. The thermometer runs down to more than sixty degrees below freezing, a temperature tolerable in calm weather, but distressing in a wind. The eye-piece of the telescope must be protected now with leather, for the skin is destroyed that comes in contact with cold metal. The voice at a mile's distance can be heard distinctly. Happy the day when first the sun is seen to graze the edge of the horizon; but summer must come, and the heat of a constant day must accumulate, and summer wane, before the ice is melted. Then the ice cracks, like cannons over-charged, and moves with a loud grinding noise. But not yet is escape to be made with safety. After a detention of ten months, Parry got free; but, in escaping, narrowly missed the destruction of both ships, by their being "nipped" between the mighty mass and the unyielding shore. What animals are found on Melville Island we may judge from the results of sport during ten months' detention. The island exceeds five thousand miles square, and yielded to the gun, three musk oxen, twenty-four deer, sixty-eight hares, fifty-three geese, fifty-nine ducks, and one hundred and forty-four ptarmigans, weighing together three thousand seven hundred and sixty-six pounds—not quite two ounces of meat per day to every man. Lichens, stunted grass, saxifrage, and a feeble willow, are the plants of Melville Island, but in sheltered nooks there are found sorrel, poppy, and a yellow buttercup. Halos and double suns are very common consequences of refraction in this quarter of the world. Franklin returned from his first and most famous voyage with his men all safe and sound, except the loss of a few fingers, frost-bitten. We sail back only as far as Regent's Inlet, being bound for Behring Strait.

The reputation of Sir John Ross being clouded by discontent expressed against his first expedition, Felix Booth, a rich distiller, provided seventeen thousand pounds to enable his friend to redeem his credit. Sir John accordingly, in 1829, went out in the *Victory*, provided with steam-machinery that did not answer well. He was accompanied by Sir James Ross, his nephew. He it was who, on this occasion, first surveyed Regent's Inlet, down which we are now sailing with our Phantom Ship. The coast on our right hand, westward, which Parry saw, is called North Somerset, but farther south, where the inlet widens, the land is named Boothia Felix. Five years before this, Parry, in his third voyage, had attempted to pass down Regent's Inlet, where among ice and storm, one of his ships, the *Hecla*, had been driven violently ashore, and of necessity abandoned. The stores had been removed, and Sir John was able now to replenish his own vessel from them. Rounding a point at the bottom of Prince Regent's Inlet, we find Felix Harbour, where Sir John Ross wintered. His nephew made from this point scientific explorations; discovered a strait, called after him the Strait of James Ross, and on the northern shore of this strait, on the main land of Boothia, planted the British flag on the Northern Magnetic Pole. The ice broke up, so did the *Victory*; after a hairbreadth escape, the party found a searching vessel and arrived home after an absence of four years and five months, Sir John Ross having lost his ship, and won his reputation, The friend in need was made a baronet for his munificence; Sir John was reimbursed for all his losses, and the crew liberally taken care of. Sir James Ross had a rod and flag signifying "Magnetic Pole," given to him for a new crest, by the Heralds' College, for which he was no doubt greatly the better.

We have sailed northward to get into Hudson Strait, the high road into Hudson Bay. Along the shore are Esquimaux in boats, extremely active, but these filthy creatures we pass by; the Esquimaux in Hudson Strait are like the negroes of the coast, demoralised by intercourse with European traders. These are not true pictures of the loving children of the north. Our "Phantom" floats on the wide waters of Hudson Bay—the grave of its discoverer. Familiar as the story is of Henry Hudson's fate, for John King's sake how gladly we repeat it. While sailing on the waters he discovered, in 1611, his men mutinied; the mutiny was aided by Henry Green, a prodigal, whom Hudson had generously shielded from ruin. Hudson, the master, and his son, with six sick or disabled members of the crew, were driven from their cabins, forced into a little shallop, and committed helpless to the water and the ice. But there was one stout man, John King, the carpenter, who stepped into the boat, abjuring his companions, and chose rather to die than even passively be partaker in so foul a crime. John King, we who live after will remember you.

Here on aim island, Charlton Island, near our entrance to the bay, in 1631, wintered poor Captain James with his wrecked crew. This is a point outside the Arctic circle, but quite cold enough. Of nights, with a good fire in the house they built, hoar frost covered their beds, and the cook's water in a metal pan before the fire was warm on one side and froze on the other. Here "it snowed and froze extremely, at which time we, looking from the shore towards the ship, she appeared a piece of ice in the fashion of a ship, or a ship resembling a piece of ice." Here the gunner, who hand lost his leg, besought that, "for the little the he had to live, he might drink sack altogether." He died and was buried in the ice far from the vessel, but when afterwards two more were dead of scurvy, and the others, in a miserable state, were working with faint hope about their shattered vessel, the gunner was found to have returned home to the old vessel; his leg had penetrated through a port-hole. They "digged him clear out, and he was as free from noisomeness," the record says, "as when we first committed him to the sea. This alteration had the ice, and water, and time, only wrought on him, that his flesh would slip up and down upon his bones, like a glove on a man's hand. In the evening we buried him by the others." These worthy souls, laid up with the agonies of scurvy, knew that in action was their only hope; they forced their limbs to labour, among ice and water, every day. They set about the building of a boat, but the hard frozen wood had broken their axes, so they made shift with the pieces. To fell a tree, it was first requisite to light in fire around it, and the carpenter could only labour with his wood over a fire, or else it was like stone under his tools. Before the boat was made they buried the carpenter. The captain exhorted them to put their trust in God; "His will be done. If it be our fortune to end our days here, we are as near Heaven as in England. They all protested to work to the utmost of their strength, and that they would refuse nothing that I should order them to do to the utmost hazard of their lives. I thanked them all." Truly the North Pole has its triumphs. If we took no account of the fields of trade opened by our Arctic explorers, if we thought nothing of the wants of science in comparison with the lives lost in supplying them, is not the loss of life a gain, which proves and tests the fortitude of noble hearts, and teaches us respect for human nature? All the lives that have been lost among these Polar regions are less in number than the dead upon a battle-field. The battle-field inflicted shame upon our race—is it with shame that our hearts throb in following these Arctic heroes? March 31st, says Captain James, "was very cold, with snow and hail, which pinched our sick men more than any time this year. This evening, being May eve, we returned late from our work to our house, and made a good fire, and chose ladies, and ceremoniously wore their names in our caps, endeavouring to revive ourselves by any means. On the 15th, I manured a little patch of ground that was bare of snow, and sowed it with pease, hoping to have some shortly to eat, for as yet we could see no green thing to comfort us." Those pease saved the party; as they came up the young shoots were boiled and eaten, so their health began to mend, and they recovered from their scurvy. Eventually, after other perils, they succeeded in making their escape.

A strait, called Sir Thomas Rowe's Welcome, leads due north out of Hudson Bay, being parted by Southampton Island from the strait through which we entered. Its name is quaint, for so was its discoverer, Luke Fox, a worthy man, addicted much to euphuism. Fox sailed from London in the same year in which James sailed from Bristol. They were rivals. Meeting in Davis Straits, Fox dined on board his friendly rival's vessel, which was very unfit for the service upon which it went. The sea washed over them and came into the cabin, so says Fox, "sauce would not have been wanted if there had been roast mutton." Luke Fox, being ice-bound and in peril, writes, "God thinks upon our imprisonment within a *supersedeas*;" but he was a good and honourable man as wall as euphuist. His "Sir Thomas Rowe's Welcome" leads into Fox Channel: our "Phantom Ship" is pushing through the welcome passes on the left-hand Repulse Bay. This portion of the Arctic regions, with Fox Channel, is extremely perilous. Here Captain Lyon, in the *Griper*, was thrown anchorless upon the mercy of a stormy sea, ice crashing around him. One island in Fox Channel is called Mill Island, from the incessant grinding of great masses of ice collected there. In the northern part of Fox Channel, on the western shore, is Melville Peninsula, where Parry wintered on his second voyage. Here let us go ashore and see a little colony of Esquimaux.

Their limits are built of blocks of snow, and arched, having an ice pane for a window. They construct their arched entrance and their hemispherical roof on the true principles of architecture. Those wise men, the Egyptians, made their arch by hewing the stones out of shape; the Esquimaux have the true secret. Here they are, with little food in winter and great appetites; devouring a whole walrus when they get it, and taking the chance of hunger for the next eight days—hungry or full, for ever happy in their lot—here are the Esquimaux. They are warmly clothed, each in a double suit of skins sewn neatly together. Some are singing, with good voices too. Please them, and they straightway dance; activity is good in a cold climate: Play to them on the flute, or if you can sing well, sing, or turn a barrel-organ, they are mute, eager with wonder and delight; their love of music is intense. Give them a pencil, and, like children, they will draw. Teach them and they will learn, oblige them and they will be grateful. "Gentle and loving savages," one of our old worthies called them, and the Portuguese were so much impressed with their teachable and gentle conduct, that a Venetian ambassador writes, "His serene majesty contemplates deriving great advantage from the country, not only on account of the timber of which he has occasion, but of the inhabitants, who are admirably calculated for labour, and are the best I have ever seen." The Esquimaux, of course, will learn vice, and in the region visited by whale ships, vice enough has certainly been taught him. Here are the dogs, who will eat old coats, or anything; and, near the dwellings, here is a snow-bunting—robin redbreast of the Arctic lands. A party of our sailors once, on landing, took some sticks from a large heap, and uncovered the nest of a snow-bunting with young, the bird flew to a little distance, but seeing that the men sat down, and harmed her not, continued to seek food and supply her little ones, with full faith in the good intentions of the party. Captain Lyon found a child's grave partly uncovered, and a snow-bunting had built its nest upon the infant's bosom.

Sailing round Melville Peninsula, we come into the Gulf of Akkolee, through Fury and Hecla Straits, discovered by Parry. So we get back to the bottom of Regent's Inlet, which we quitted a short time ago, and sailing in the neighbourhood of the magnetic pole, we reach the estuary of Back's River, on the north-east coast of America. We pass then through a strait, discovered in 1839 by Dean and Simpson, still coasting along the northern shore of America, on the great Stinking Lake, as Indians call this ocean. Boats, ice permitting, and our "Phantom Ship," of course, can coast all the way to Behring Strait. The whole coast has been explored by Sir John Franklin, Sir John Richardson, and Sir George Back, who have earned their knighthoods through great peril. As we pass Coronation Gulf—the scene of Franklin, Richardson, and Back's first exploration from the Coppermine River—we revert to the romantic story of their journey back, over a land of snow and frost, subsisting upon lichens, with companions starved to death, where they plucked wild leaves for tea, and ate their shoes for supper; the tragedy by the river; the murder of poor Hood, with a book of prayers in his hand; Franklin at Fort Enterprise, with two companions at the point of death, himself gaunt, hollow-eyed, feeding on pounded bones, raked from the dunghill; the arrival of Dr. Richardson and the brave sailor; their awful story of the cannibal Michel;—we revert to these things with a shudder. But we must continue on our route. The current still flows westward, bearing now large quantities of driftwood out of the Mackenzie River. At the name of Sir Alexander Mackenzie, also, we might pause, and talk over the bold achievements of another Arctic hero; but we pass on, by a rugged and inhospitable coast, unfit for vessels of large draught—pass the broad mouth of the Youcon, pass Point Barrow, Icy Cape, and are in Behring Strait. Had we passed on, we should have found the Russian Arctic coast line, traced out by a series of Russian explorers; of whom the most illustrious—Baron Von Wrangell—states, that beyond a certain distance to the northward there is always found what he calls the *Polynja* (open water). This is the fact adduced by those who adhere to the old fancy that there is a sea about the Pole itself quite free from ice.

We pass through Behring Straits. Behring, a Dane by birth, but in the Russian service, died here in 1741, upon the scene of his discovery. He and his crew, victims of scurvy, were unable to manage their vessel in a storm; and it was at length wrecked on a barren island, there, where "want, nakedness, cold, sickness, impatience, and despair, were their daily guests," Behring, his lieutenant, and the master died.

Now we must put a girdle round the world, and do it with the speed of Ariel. Here we are already in the heats of the equator. We can do no more than remark, that if air and water are heated at the equator, and frozen at the poles, there will be equilibrium destroyed, and constant currents caused. And so it happens, so we get the prevailing winds, and all the currents of the ocean. Of these, some of the uses, but by no means all, are obvious. We urge our "Phantom" fleetly to the southern pole. Here, over the other hemisphere of the earth, there shines another hemisphere of heaven. The stars are changed; the southern cross, the Magellanic clouds, the "coal-sack" in the milky way, attract our notice. Now we are in the southern latitude that corresponds to England in the north; nay, at a greater distance from the Pole, we find Kerguelen's Land, emphatically called "The Isle of Desolation." Icebergs float much further into the warm sea on this side of the equator before they dissolve. The South Pole is evidently a more thorough refrigerator than the North. Why is this? We shall soon see. We push through pack-ice, and through floes and fields, by lofty bergs, by an island or two covered with penguins, until there lies before us a long range of mountains, nine or ten thousand feet in height, and all clad in eternal snow. That is a portion of the Southern Continent. Lieutenant Wilkes, in the American exploring expedition, first discovered this, and mapped out some part of the coast, putting a few clouds in likewise—a mistake easily made by those who omit to verify every foot of land. Sir James Ross, in his most successful South Pole Expedition, during the years 1839-43, sailed over some of this land, and confirmed the rest. The Antarctic, as well as the Arctic honours he secured for England, by turning a corner of the land, and sailing far southward, along an impenetrable icy barrier, to the latitude of seventy-eight degrees, nine minutes. It is an elevated continent, with many lofty ranges. On the extreme southern point reached by the ships, a magnificent volcano was seen spouting fire and smoke out of the everlasting snow. This volcano, twelve thousand four hundred feet high, was named Mount Erebus; for the *Erebus* and *Terror* long sought anxiously among the bays, and sounds, and creeks of the North Pole, then coasted by the solid ice walls of the south.

H. M.

A DISCOURSE WRITTEN BY SIR HUMPHREY GILBERT, KNIGHT.

To prove a Passage by the North-West to Cathay and the East Indies.

CHAPTER I.
TO PROVE BY AUTHORITY A PASSAGE TO BE ON THE NORTH SIDE OF AMERICA, TO GO TO CATHAY AND THE EAST INDIES.

When I gave myself to the study of geography, after I had perused and diligently scanned the descriptions of Europe, Asia, and Africa, and conferred them with the maps and globes both antique and modern, I came in fine to the fourth part of the world, commonly called America, which by all descriptions I found to be an island environed round about with the sea, having on the south side of it the Strait of Magellan, on the west side the Mare de Sur, which sea runneth towards the north, separating it from the east parts of Asia, where the dominions of the Cathaians are. On the east part our west ocean, and on the north side the sea that severeth it from Greenland, through which northern seas the passage lieth, which I take now in hand to discover.

Plato in his *Timaeus* and in the dialogue called *Critias*, discourses of an incomparable great island then called Atlantis, being greater than all Africa and Asia, which lay westward from the Straits of Gibraltar, navigable round about: affirming, also, that the princes of Atlantis did as well enjoy the governance of all Africa and the most part of Europe as of Atlantis itself.

Also to prove Plato's opinion of this island, and the inhabiting of it in ancient time by them of Europe, to be of the more credit: Marinæus Siculus, in his Chronicle of Spain, reporteth that there hath been found by the Spaniards in the gold mines of America certain pieces of money, engraved with the image of Augustus Cæsar; which pieces were sent to the Pope for a testimony of the matter by John Rufus, Archbishop of Constantinum.

Moreover, this was not only thought of Plato, but by Marsilius Ficinus, an excellent Florentine philosopher, Crantor the Grecian, Proclus, also Philo the famous Jew (as appeareth in his book *De Mundo*, and in the Commentaries upon Plato), to be overflown, and swallowed up with water, by reason of a mighty earthquake and streaming down of the heavenly flood gates. The like thereof happened unto some part of Italy, when by the forcibleness of the sea, called Superum, it cut off Sicily from the continent of Calabria, as appeareth in Justin in the beginning of his fourth book. Also there chanced the like in Zeeland, a part of Flanders.

And also the cities of Pyrrha and Antissa, about Palus Meotis; and also the city Burys, in the Corinthian Gulf, commonly called Sinus Corinthiacus, have been swallowed up with the sea, and are not at this day to be discerned: by which accident America grew to be unknown, of long time, unto us of the later ages, and was lately discovered again by Americus Vespucius, in the year of our Lord 1497, which some say to have been first discovered by Christopher Columbus, a Genoese, Anno 1492.

The same calamity happened unto this isle of Atlantis six hundred and odd years before Plato's time, which some of the people of the south-east parts of the world accounted as nine thousand years; for the manner then was to reckon the moon's period of the Zodiac for a year, which is our usual month, depending a Luminari minore.

So that in these our days there can no other main or island be found or judged to be parcel of this Atlantis than those western islands, which now bear the name of America; countervailing thereby the name of Atlantis in the knowledge of our age.

Then, if when no part of the said Atlantis was oppressed by water and earthquake, the coasts round about the same were navigable, a far greater hope now remaineth of the same by the north-west, seeing the most part of it was since that time swallowed up with water, which could not utterly take away the old deeps and channels, but, rather, be many occasion of the enlarging of the old, and also an enforcing of a great many new; why then should we now doubt of our North-West Passage and navigation from England to India, etc., seeing that Atlantis, now called America, was ever known to be an island, and in those days navigable round about, which by access of more water could not be diminished?

Also Aristotle in his book *De Mundo*, and the learned German, Simon Gryneus, in his annotations upon the same, saith that the whole earth (meaning thereby, as manifestly doth appear, Asia, Africa, and Europe, being all the countries then known) to be but one island, compassed about with the reach of the Atlantic sea; which likewise approveth America to be an island, and in no part adjoining to Asia or the rest.

Also many ancient writers, as Strabo and others, called both the ocean sea (which lieth east of India) Atlanticum Pelagus, and that sea also on the west coasts of Spain and Africa, Mare Atlanticum; the distance between the two coasts is almost half the compass of the earth.

So that it is incredible, as by Plato appeareth manifestly, that the East Indian Sea had the name of Atlanticum Pelagus, of the mountain Atlas in Africa, or yet the sea adjoining to Africa had name Oceanus Atlanticus, of the same mountain; but that those seas and the mountain Atlas were so called of this great island Atlantis, and that the one and the other had their names for a memorial of the mighty Prince Atlas, sometime king thereof, who was Japhet, youngest son to Noah, in whose time the whole earth was divided between the three brethren, Shem, Ham, and Japhet.

Wherefore I am of opinion that America by the north-west will be found favourable to this our enterprise, and am the rather emboldened to believe the same, for that I find it not only confirmed by Plato, Aristotle, and other ancient philosophers, but also by the best modern geographers, as Gemma Frisius, Munsterus, Appianus Hunterus, Gastaldus, Guyccardinus, Michael Tramesinus, Franciscus Demongenitus, Barnardus, Puteanus, Andreas Vavasor, Tramontanus, Petrus Martyr, and also Ortelius, who doth coast out in his general map (set out Anno 1569) all the countries and capes on the north-west side of America from Hochelega to Cape de Paramantia, describing likewise the sea-coasts of Cathay and Greenland, towards any part of America, making both Greenland and America islands disjoined by a great sea from any part of Asia.

All which learned men and painful travellers have affirmed with one consent and voice, that America was an island, and that there lieth a great sea between it, Cathay, and Greenland, by the which any man of our country that will give the attempt, may with small danger pass to Cathay, the Moluccas, India, and all other places in the east in much shorter time than either the Spaniard or Portuguese doth, or may do, from the nearest part of any of their countries within Europe.

What moved these learned men to affirm thus much I know not, or to what end so many and sundry travellers of both ages have allowed the same; but I conjecture that they would never have so constantly affirmed, or notified their opinions therein to the world, if they had not had great good cause, and many probable reasons to have led them thereunto.

Now lest you should make small account of ancient writers or of their experiences which travelled long before our times, reckoning their authority amongst fables of no importance, I have for the better assurance of those proofs set down some part of a discourse, written in the Saxon tongue, and translated into English by Master Noel, servant to Master Secretary Cecil, wherein there is described a navigation which one other made, in the time of King Alfred, King of Wessex, Anne 871, the words of which discourse were these: "He sailed right north, having always the desert land on the starboard, and on the larboard the main sea, continuing his course, until he perceived that the coast bowed directly towards the east or else the sea opened into the land he could not tell how far, where he was compelled to stay until he had a western wind or somewhat upon the north, and sailed thence directly east along the coast, so far as he was able in four days, where he was again enforced to tarry until he had a north wind, because the coast there bowed directly towards the south, or at least opened he knew not how far into the land, so that he sailed thence along the coast continually full south, so far as he could travel in the space of five days, where he discovered a mighty river which opened far into the land, and in the entry of this river he turned back again."

Whereby it appeareth that he went the very way that we now do yearly trade by S. Nicholas into Muscovia, which way no man in our age knew for certainty to be sea, until it was since discovered by our Englishmen in the time of King Edward I., but thought before that time that Greenland had joined to Normoria Byarmia, and therefore was accounted a new discovery, being nothing so indeed, as by this discourse of Ochther's it appeareth.

Nevertheless if any man should have taken this voyage in hand by the encouragement of this only author, he should have been thought but simple, considering that this navigation was written so many years past, in so barbarous a tongue by one only obscure author, and yet we in these our days find by our own experiences his former reports to be true.

How much more, then, ought we to believe this passage to Cathay to be, being verified by the opinions of all the best, both antique and modern geographers, and plainly set out in the best and most allowed maps, charts, globes, cosmographical tables, and discourses of this our age and by the rest not denied, but left as a matter doubtful.

CHAPTER II.

1. All seas are maintained by the abundance of water, so that the nearer the end any river, bay, or haven is, the shallower it waxeth (although by some accidental bar it is sometime found otherwise), but the farther you sail west from Iceland, towards the place where this strait is thought to be, the more deep are the seas, which giveth us good hope of continuance of the same sea, with Mare del Sur, by some strait that lieth between America, Greenland, and Cathay.

2. Also, if that America were not an island, but a part of the continent adjoining to Asia, either the people which inhabit Mangia, Anian, and Quinzay, etc., being borderers upon it, would before this time have made some road into it, hoping to have found some like commodities to their own.

3. Or else the Syrians and Tartars (which oftentimes heretofore have sought far and near for new seats, driven thereunto through the necessity of their cold and miserable countries) would in all this time have found the way to America and entered the same had the passages been never so strait or difficult, the country being so temperate, pleasant, and fruitful in comparison of their own. But there was never any such people found there by any of the

Spaniards, Portuguese, or Frenchmen, who first discovered the inland of that country, which Spaniards or Frenchmen must then of necessity have seen some one civilised man in America, considering how full of civilised people Asia is; but they never saw so much as one token or sign that ever any man of the known part of the world had been there.

4. Furthermore, it is to be thought, that if by reason of mountains or other craggy places the people neither of Cathay or Tartary could enter the country of America, or they of America have entered Asia if it were so joined, yet some one savage or wandering-beast would in so many years have passed into it; but there hath not any time been found any of the beasts proper to Cathay or Tartary, etc., in America; nor of those proper to America in Tartary, Cathay, etc., or in any part of Asia, which thing proveth America not only to be one island, and in no part adjoining to Asia, but also that the people of those countries have not had any traffic with each other.

5. Moreover at the least some one of those painful travellers which of purpose have passed the confines of both countries, with intent only to discover, would, as it is most likely, have gone from the one to the other, if there had been any piece of land, or isthmus, to have joined them together, or else have declared some cause to the contrary.

6. But neither Paulus Venetus, who lived and dwelt a long time in Cathay, ever came into America, and yet was at the sea coasts of Mangia over against it, where he was embarked and performed a great navigation along those seas; neither yet Veratzanus or Franciscus Vasquez de Coronado, who travelled the north part of America by land, ever found entry from thence by land to Cathay, or any part of Asia.

7. Also it appeareth to be an island, insomuch as the sea runneth by nature circularly from the east to the west, following the diurnal motion of the *Primum Mobile*, and carrieth with it all inferior bodies movable, as well celestial as elemental; which motion of the waters is most evidently seen in the sea, which lieth on the south side of Africa, where the current that runneth from the east to the west is so strong (by reason of such motion) that the Portuguese in their voyages eastward to Calicut, in passing by the Cape of Good Hope, are enforced to make divers courses, the current there being so swift, as it striketh from thence, all along westward, upon the straits of Magellan, being distant from thence near the fourth part of the longitude of the earth: and not having free passage and entrance through that frith towards the west, by reason of the narrowness of the said strait of Magellan, it runneth to salve this wrong (Nature not yielding to accidental restraints) all along the eastern coasts of America northwards so far as Cape Frido, being the farthest known place of the same continent towards the north, which is

about four thousand eight-hundred leagues, reckoning therewithal the trending of the land.

8. So that this current, being continually maintained with such force as Jacques Cartier affirmeth it to be, who met with the same, being at Baccalaos as he sailed along the coasts of America, then, either it must of necessity have way to pass from Cape Frido through this frith, westward towards Cathay, being known to come so far only to salve his former wrongs by the authority before named; or else it must needs strike over upon the coast of Iceland, Lapland, Finmark, and Norway (which are east from the said place about three hundred and sixty leagues) with greater force than it did from the Cape of Good Hope upon the strait of Magellan, or from the strait of Magellan to Cape Frido; upon which coasts Jacques Cartier met with the same, considering the shortness of the cut from the said Cape Frido to Iceland, Lapland, etc. And so the cause efficient remaining, it would have continually followed along our coasts through the narrow seas, which it doeth not, but is digested about the north of Labrador by some through passage there through this frith.

The like course of the water, in some respect, happeneth in the Mediterranean Sea (as affirmeth Contorenus), where, as the current which cometh from Tanais and the Euxine, running along all the coasts of Greece, Italy, France, and Spain, and not finding sufficient way out through Gibraltar by means of the straitness of the frith, it runneth back again along the coasts of Barbary by Alexandria, Natolia, etc.

It may, peradventure, be thought that this course of the sea doth sometime surcease and thereby impugn this principle, because it is not discerned all along the coast of America in such sort as Jacques Cartier found it, whereunto I answer this: That albeit in every part of the coast of America or elsewhere this current is not sensibly perceived, yet it hath evermore such like motion, either the uppermost or nethermost part of the sea; as it may be proved true, if you sink a sail by a couple of ropes near the ground, fastening to the nethermost corners two gun chambers or other weights, by the driving whereof you shall plainly perceive the course of the water and current running with such like course in the bottom. By the like experiment you may find the ordinary motion of the sea in the ocean, how far soever you be off the land.

9. Also, there cometh another current from out the north-east from the Scythian Sea (as Master Jenkinson, a man of rare virtue, great travel, and experience, told me), which runneth westward towards Labrador, as the other did which cometh from the south; so that both these currents must have way through this our strait, or else encounter together and run contrary courses in one line, but no such conflicts of streams or contrary courses are

found about any part of Labrador or Newfoundland, as witness our yearly fishers and other sailors that way, but is there separated as aforesaid, and found by the experience of Barnarde de la Torre to fall into Mare del Sur.

10. Furthermore, the current in the great ocean could not have been maintained to run continually one way from the beginning of the world unto this day, had there not been some through passage by the strait aforesaid, and so by circular motion be brought again to maintain itself, for the tides and courses of the sea are maintained by their interchangeable motions, as fresh rivers are by springs, by ebbing and flowing, by rarefaction and condensation.

So that it resteth not possible (so far as my simple reason can comprehend) that this perpetual current can by any means be maintained, but only by a continual reaccess of the same water, which passeth through the strait, and is brought about thither again by such circular motion as aforesaid, and the certain falling thereof by this strait into Mare del Sur is proved by the testimony and experience of Barnarde de la Torre, who was sent from P. de la Natividad to the Moluccas, 1542, by commandment of Anthony Mendoza, then Viceroy of Nova Hispania, which Barnarde sailed 750 leagues on the north side of the Equator, and there met with a current which came from the north-east, the which drove him back again to Tidore.

Wherefore this current being proved to come from the Cape of Good Hope to the strait of Magellan, and wanting sufficient entrance there, is by the necessity of Nature's force brought to Terra de Labrador, where Jacques Cartier met the same, and thence certainly known not to strike over upon Iceland, Lapland, etc., and found by Barnarde de la Torre, in Mare del Sur, on the backside of America, therefore this current, having none other passage, must of necessity fall out through this strait into Mare del Sur, and so trending by the Moluccas, China, and the Cape of Good Hope, maintaineth itself by circular motion, which is all one in Nature with motus ab oriente in occidentem.

So that it seemeth we have now more occasion to doubt of our return than whether there be a passage that way, yea or no: which doubt hereafter shall be sufficiently removed; wherefore, in my opinion reason itself grounded upon experience assureth us of this passage if there were nothing else to put us in hope thereof. But lest these might not suffice, I have added in this chapter following some further proof thereof, by the experience of such as have passed some part of this discovery, and in the next adjoining to that the authority of those which have sailed wholly through every part thereof.

CHAPTER III.
TO PROVE BY EXPERIENCE OF SUNDRY MEN'S TRAVELS THE OPENING OF SOME PART OF THIS NORTH-WEST PASSAGE, WHEREBY GOOD HOPE REMAINETH OF THE REST.

1. Paulus Venetus, who dwelt many years in Cathay, affirmed that he had sailed 1,500 miles upon the coast of Mangia and Anian, towards the north-east, always finding the seas open before him, not only as far as he went, but also as far as he could discern.

2. Also Franciscus Vasquez de Coronado, passing from Mexico by Cevola, through the country of Quiver to Sierra Nevada, found there a great sea, where were certain ships laden with merchandise, the mariners wearing on their heads the pictures of certain birds called Alcatrarzi, part whereof were made of gold and part of silver; who signified by signs that they were thirty days coming thither, which likewise proveth America by experience to be disjoined from Cathay, on that part, by a great sea, because they could not come from any part of America as natives thereof; for that, so far as is discovered, there hath not been found there any one ship of that country.

3. In like manner, Johann Baros testifieth that the cosmographers of China (where he himself had been) affirm that the sea coast trendeth from thence north-east to fifty degrees of septentrional latitude, being the farthest part that way, which the Portuguese had then knowledge of; and that the said cosmographers knew no cause to the contrary, but that it might continue farther.

By whose experiences America is proved to be separate from those parts of Asia, directly against the same. And not contented with the judgments of these learned men only, I have searched what might be further said for the confirmation hereof.

4. And I found that Franciscus Lopez de Gomara affirmeth America to be an island, and likewise Greenland; and that Greenland is distant from Lapland forty leagues, and from Terra de Labrador fifty.

5. Moreover Alvarez Nunmius, a Spaniard, and learned cosmographer, and Jacques Cartier, who made two voyages into those parts, and sailed five hundred miles upon the north-east coasts of America.

6. Likewise Hieronimus Fracastorius, a learned Italian, and traveller in the north parts of the same land.

7. Also Jacques Cartier, having done the like, heard say at Hochelaga, in Nova Francia, how that there was a great sea at Saguinay, whereof the end was not known: which they presupposed to be the passage to

Cathay. Furthermore, Sebastian Cabot, by his personal experience and travel, has set forth and described this passage in his charts which are yet to be seen in the Queen's Majesty's Privy Gallery at Whitehall, who was sent to make this discovery by King Henry VII. and entered the same straits, affirming that he sailed very far westward with a quarter of the north, on the north side of Terra de Labrador, the 11th of June, until he came to the septentrional latitude of sixty-seven and a half degrees, and finding the seas still open, said, that he might and would have gone to Cathay if the mutiny of the master and mariners had not been.

Now, as these men's experience have proved some part of this passage, so the chapter following shall put you in full assurance of the rest by their experiences which have passed through every part thereof.

CHAPTER IV.
TO PROVE BY CIRCUMSTANCE THAT THE NORTH-WEST PASSAGE HATH BEEN SAILED THROUGHOUT.

The diversity between brute beasts and men, or between the wise and the simple, is, that the one judgeth by sense only, and gathereth no surety of anything that he hath not seen, felt, heard, tasted, or smelled: and the other not so only, but also findeth the certainty of things, by reason, before they happen to be tried, wherefore I have added proofs of both sorts, that the one and the other might thereby be satisfied.

1. First, as Gemma Frisius reciteth, there went from Europe three brethren though this passage: whereof it took the name of Fretum trium fratrum.

2. Also Pliny affirmeth out of Cornelius Nepos (who wrote fifty-seven years before Christ) that there were certain Indians driven by tempest upon the coast of Germany which were presented by the King of Suevia unto Quintus Metellus Celer, then Pro-Consul of France.

3. And Pliny upon the same saith that it is no marvel, though there be sea by the north, where there is such abundance of moisture; which argueth, that he doubted not of a navigable passage that way, through which those Indians came.

4. And for the better proof that the same authority of Cornelius Nepos is not by me wrested to prove my opinion of the North-West Passage, you shall find the same affirmed more plainly in that behalf by the excellent geographer Dominicus Marius Niger, who showeth how many ways the Indian sea stretcheth itself, making in that place recital of certain Indians that were likewise driven through the north seas from India, upon the coasts of Germany, by great tempest, as they were sailing in trade of merchandise.

5. Also, whiles Frederick Barbarossa reigned Emperor, A.D. 1160, there came certain other Indians upon the coast of Germany.

6. Likewise Othon, in the story of the Goths, affirmeth that in the time of the German Emperors there were also certain Indians cast by force of weather upon the coast of the said country, which foresaid Indians could not possibly have come by the south-east, south-west, nor from any part of Africa or America, nor yet by the north-east: therefore they came of necessity by this our North-West Passage.

CHAPTER V.
TO PROVE THAT THESE INDIANS, AFORENAMED, CAME NOT BY THE SOUTH-EAST, SOUTH-WEST, NOR FROM ANY OTHER PART OF AFRICA OR AMERICA.

1. They could not come from the south-east by the Cape of Good Hope, because the roughness of the seas there is such—occasioned by the currents and great winds in that part—that the greatest armadas the King of Portugal hath cannot without great difficulty pass that way, much less, then, a canoe of India could live in those outrageous seas without shipwreck, being a vessel but of very small burden, and the Indians have conducted themselves to the place aforesaid, being men unexpert in the art of navigation.

2. Also, it appeareth plainly that they were not able to come from along the coast of Africa aforesaid to those parts of Europe, because the winds do, for the most part, blow there easterly or from the shore, and the current running that way in like sort, would have driven them westward upon some part of America, for such winds and tides could never have led them from thence to the said place where they were found, nor yet could they have come from any of the countries aforesaid, keeping the seas always, without skilful mariners to have conducted them such like courses as were necessary to perform such a voyage.

3. Presupposing also, if they had been driven to the west, as they must have been, coming that way, then they should have perished, wanting supply of victuals, not having any place—once leaving the coast of Africa—until they came to America, north of America, until they arrived upon some part of Europe or the islands adjoining to it to have refreshed themselves.

4. Also, if, notwithstanding such impossibilities, they might have recovered Germany by coming from India by the south-east, yet must they without all doubt have struck upon some other part of Europe before their arrival there, as the isles of Madeira, Portugal, Spain, France, England, Ireland, etc., which, if they had done, it is not credible that they should or would have departed undiscovered of the inhabitants; but there was never found in those days any such ship or men, but only upon the coasts of Germany, where they have

been sundry times and in sundry ages cast ashore; neither is it like that they would have committed themselves again to sea, if they had so arrived, not knowing where they were, nor whither to have gone.

5. And by the south-west it is impossible, because the current aforesaid, which cometh from the east, striketh with such force upon the Straits of Magellan, and falleth with such swiftness and fury into Mare de Sur, that hardly any ship—but not possibly a canoe, with such unskilful mariners—can come into our western ocean through that strait from the west seas of America, as Magellan's experience hath partly taught us.

6. And further, to prove that these people so arriving upon the coast of Germany were Indians, and not inhabiters of any part either of Africa or America, it is manifest, because the natives, both of Africa and America, neither had, or have at this day, as is reported, other kind of boats than such as do bear neither masts nor sails, except only upon the coasts of Barbary and the Turks' ships, but do carry themselves from place to place near the shore by the oar only.

CHAPTER VI.
TO PROVE THAT THOSE INDIANS CAME NOT BY THE NORTH-EAST, AND THAT THERE IS NO THROUGH NAVIGABLE PASSAGE THAT WAY.

1. It is likely that there should be no through passage by the north-east whereby to go round about the world, because all seas, as aforesaid, are maintained by the abundance of water, waxing more shallow and shelving towards the end, as we find it doth, by experience, in the Frozen Sea, towards the east, which breedeth small hope of any great continuance of that sea to be navigable towards the east, sufficient to sail thereby round about the world.

2. Also, it standeth scarcely with reason that the Indians dwelling under the Torrid Zone could endure the injury of the cold air, about the northern latitude of 80 degrees, under which elevation the passage by the north-east cannot be, as the often experiences had of all the south part of it showeth, seeing that some of the inhabitants of this cold climate, whose summer is to them an extreme winter, have been stricken to death with the cold damps of the air, about 72 degrees, by an accidental mishap, and yet the air in such like elevation is always cold, and too cold for such as the Indians are.

3. Furthermore, the piercing cold of the gross thick air so near the Pole will so stiffen the sails and ship tackling, that no mariner can either hoist or strike them—as our experience, far nearer the south than this passage is presupposed to be, hath taught us—without the use whereof no voyage can be performed.

4. Also, the air is so darkened with continual mists and fogs so near the Pole, that no man can well see either to guide his ship or to direct his course.

5. Also the compass at such elevation doth very suddenly vary, which things must of force have been their destruction, although they had been men of much more skill than the Indians are.

6. Moreover, all bays, gulfs, and rivers do receive their increase upon the flood, sensibly to be discerned on the one side of the shore or the other, as many ways as they be open to any main sea, as the Mediterranean, the Red Sea, the Persian Gulf, Sinus Bodicus, the Thames, and all other known havens or rivers in any part of the world, and each of them opening but on one part to the main sea, do likewise receive their increase upon the flood the same way, and none other, which the Frozen Sea doth, only by the west, as Master Jenkinson affirmed unto me, and therefore it followeth that this north-east sea, receiving increase only from the west, cannot possibly open to the main ocean by the east.

7. Moreover, the farther you pass into any sea towards the end of it, of that part which is shut up from the main sea, as in all those above-mentioned, the less and less the tides rise and fall. The like whereof also happeneth in the Frozen Sea, which proveth but small continuance of that sea toward the east.

8. Also, the farther ye go towards the east in the Frozen Sea the less soft the water is, which could not happen if it were open to the salt sea towards the east, as it is to the west only, seeing everything naturally engendereth his like, and then must it be like salt throughout, as all the seas are in such like climate and elevation. And therefore it seemeth that this north-east sea is maintained by the river Ob, and such like freshets as the Pontic Sea and Mediterranean Sea, in the uppermost parts thereof by the river Nile, the Danube, Dnieper, Tanais, etc.

9. Furthermore, if there were any such sea at that elevation, of like it should be always frozen throughout—there being no tides to hinder it—because the extreme coldness of the air in the uppermost part, and the extreme coldness of the earth in the bottom, the sea there being but of small depth, whereby the one accidental coldness doth meet with the other; and the sun, not having his reflection so near the Pole, but at very blunt angles, it can never be dissolved after it is frozen, notwithstanding the great length of their day: for that the sun hath no heat at all in his light or beams, but proceeding only by an accidental reflection which there wanteth in effect.

10. And yet if the sun were of sufficient force in that elevation to prevail against this ice, yet must it be broken before it can be dissolved, which cannot be but through the long continue of the sun above their horizon, and by that time the summer would be so far spent, and so great darkness and cold ensue,

that no man could be able to endure so cold, dark, and discomfortable a navigation, if it were possible for him then and there to live.

11. Further, the ice being once broken, it must of force so drive with the winds and tides that no ship can sail in those seas, seeing our fishers of Iceland and Newfoundland are subject to danger through the great islands of ice which fleet in the seas, far to the south of that presupposed passage.

12. And it cannot be that this North-East Passage should be any nearer the south than before recited, for then it should cut off Ciremissi and Turbi, Tartarii, with Vzesucani, Chisani, and others from the continent of Asia, which are known to be adjoining to Scythia, Tartary, etc., with the other part of the same continent.

And if there were any through passage by the north-east, yet were it to small end and purpose for our traffic, because no ship of great burden can navigate in so shallow a sea, and ships of small burden are very unfit and unprofitable, especially towards the blustering north, to perform such a voyage.

CHAPTER VII.
TO PROVE THAT THE INDIANS AFORENAMED CAME ONLY BY THE NORTH-WEST, WHICH INDUCETH A CERTAINTY OF OUR PASSAGE BY EXPERIENCE.

It is as likely that they came by the north-west as it is unlikely that they should come either by the south-east, south-west, north-east, or from any other part of Africa or America, and therefore this North-West Passage, having been already so many ways proved by disproving of the others, etc., I shall the less need in this place to use many words otherwise than to conclude in this sort, that they came only by the north-west from England, having these many reasons to lead me thereunto.

1. First, the one-half of the winds of the compass might bring them by the north-west, veering always between two sheets, with which kind of sailing the Indians are only acquainted, not having any use of a bow line or quarter wind, without the which no ship can possibly come, either by the south-east, south-west, or north-east, having so many sundry capes to double, whereunto are required such change and shifts of winds.

2. And it seemeth likely that they should come by the north-west, because the coast whereon they were driven lay east from this our passage, and all winds do naturally drive a ship to an opposite point from whence it bloweth, not being otherwise guided by art, which the Indians do utterly want, and therefore it seemeth that they came directly through this, our strait, which they might do with one wind.

3. For if they had come by the Cape of Good Hope, then must they, as aforesaid, have fallen upon the south parts of America.

4. And if by the Strait of Magellan, then upon the coasts of Africa, Spain, Portugal, France, Ireland, or England.

5. And if by the north-east, then upon the coasts of Ciremissi, Tartarii, Lapland, Iceland, Labrador, etc., and upon these coasts, as aforesaid, they have never been found.

So that by all likelihood they could never have come without shipwreck upon the coasts of Germany, if they had first struck upon the coasts of so many countries, wanting both art and shipping to make orderly discovery, and altogether ignorant both of the art of navigation and also of the rocks, flats, sands, or havens of those parts of the world, which in most of these places are plentiful.

6. And further, it seemeth very likely that the inhabitants of the most part of those countries, by which they must have come any other way besides by the north-west, being for the most part anthropophagi, or men-eaters, would have devoured them, slain them, or, at the leastwise, kept them as wonders for the gaze.

So that it plainly appeareth that those Indians—which, as you have heard, in sundry ages were driven by tempest upon the shore of Germany—came only through our North-West Passage.

7. Moreover, the passage is certainly proved by a navigation that a Portuguese made, who passed through this strait, giving name to a promontory far within the same, calling it after his own name, Promontorium Corterialis, near adjoining unto Polisacus Fluvius.

8. Also one Scolmus, a Dane, entered and passed a great part thereof.

9. Also there was one Salva Terra, a gentleman of Victoria in Spain, that came by chance out of the West Indies into Ireland, Anno 1568, who affirmed the North-West Passage from us to Cathay, constantly to be believed in America navigable; and further said, in the presence of Sir Henry Sidney, then Lord Deputy of Ireland, in my hearing, that a friar of Mexico, called Andre Urdaneta, more than eight years before his then coming into Ireland, told him there that he came from Mare del Sur into Germany through this North-West Passage, and showed Salva Terra—at that time being then with him in Mexico—a sea-card made by his own experience and travel in that voyage, wherein was plainly set down and described this North-West Passage, agreeing in all points with Ortelius' map.

And further this friar told the King of Portugal (as he returned by that country homeward) that there was of certainty such a passage north-west from England, and that he meant to publish the same; which done, the king most earnestly desired him not in any wise to disclose or make the passage known to any nation. For that (said the king) *if England had knowledge and experience thereof, it would greatly hinder both the King of Spain and me.* This friar (as Salva Terra reported) was the greatest discoverer by sea that hath been in our age. Also Salva Terra, being persuaded of this passage by the friar Urdaneta, and by the common opinion of the Spaniards inhabiting America, offered most willingly to accompany me in this discovery, which of like he would not have done if he had stood in doubt thereof.

And now, as these modern experiences cannot be impugned, so, least it might be objected that these things (gathered out of ancient writers, which wrote so many years past) might serve little to prove this passage by the north of America, because both America and India were to them then utterly unknown; to remove this doubt, let this suffice, that Aristotle (who was 300 years before Christ) named the Indian Sea. Also Berosus (who lived 330 before Christ) hath these words, *Ganges in India.*

Also in the first chapter of Esther be these words: "In the days of Ahasuerus, which ruled from India to Ethiopia," which Ahasuerus lived 580 years before Christ. Also Quintus Curtius, where he speaketh of the Conquest of Alexander, mentioneth India. Also Arianus Philostratus, and Sidrach, in his discourses of the wars of the King of Bactria, and of Garaab, who had the most part of India under his government. All which assumeth us that both India and Indians were known in those days.

These things considered, we may, in my opinion, not only assure ourselves of this passage by the north-west, but also that it is navigable both to come and go, as hath been proved in part and in all by the experience of divers as Sebastian Cabot, Corterialis, the three brethren above named, the Indians, and Urdaneta, the friar of Mexico, etc.

And yet, notwithstanding all which, there be some that have a better hope of this passage to Cathay by the north-east than by the west, whose reasons, with my several answers, ensue in the chapter following.

CHAPTER VIII.
CERTAIN REASONS ALLEGED FOR THE PROVING OF A PASSAGE BY THE NORTH-EAST BEFORE THE QUEEN'S MAJESTY, AND CERTAIN LORDS OF THE COUNCIL, BY MASTER ANTHONY JENKINSON, WITH MY SEVERAL ANSWERS THEN USED TO THE SAME.

Because you may understand as well those things alleged against me as what doth serve for my purpose, I have here added the reasons of Master Anthony Jenkinson, a worthy gentleman, and a great traveller, who conceived a better hope of the passage to Cathay from us to be by the north-east than by the north-west.

He first said that he thought not to the contrary but that there was a passage by the north-west, according to mime opinion, but he was assured that there might be found a navigable passage by the north-east from England to go to all the east parts of the world, which he endeavoured to prove three ways.

The first was, that he heard a fisherman of Tartary say in hunting the morse, that he sailed very far towards the south-east, finding no end of the sea, whereby he hoped a through passage to be that way.

Whereunto I answered that the Tartars were a barbarous people, and utterly ignorant in the art of navigation, not knowing the use of the sea-card, compass, or star, which he confessed true; and therefore they could not (said I) certainly know the south-east from the north-east in a wide sea, and a place unknown from the sight of the land.

Or if he sailed anything near the shore, yet he, being ignorant, might be deceived by the doubling of many points and capes, and by the trending of the land, albeit he kept continually along the shore.

And further, it might be that the poor fisherman through simplicity thought that there was nothing that way but sea, because he saw mine land, which proof (under correction) giveth small assurance of a navigable sea by the north-east to go round about the world, for that he judged by the eye only, seeing we in this clear air do account twenty miles a ken at sea.

His second reason is, that there was an unicorn's horn found upon the coast of Tartary, which could not come (said he) thither by any other means than with the tides, through some strait in the north-east of the Frozen Sea, there being no unicorns in any part of Asia, saving in India and Cathay, which reason, in my simple judgment, has as little force.

First, it is doubtful whether those barbarous Tartars do know an unicorn's horn, yea or no; and if it were one, yet it is not credible that the sea could have driven it so far, it being of such nature that it cannot float.

Also the tides running to and fro would have driven it as far back with the ebb as it brought it forward with the flood.

There is also a beast called Asinus Indicus (whose horn most like it was), which hath but one horn like an unicorn in his forehead, whereof there is great plenty in all the north parts thereunto adjoining, as in Lapland, Norway, Finmark, etc., as Jocobus Zeiglerus writeth in his history of Scondia.

And as Albertus saith, there is a fish which hath but one horn in his forehead like to an unicorn, and therefore it seemeth very doubtful both from whence it came, and whether it were an unicorn's horn, yea or no.

His third and last reason was, that there came a continual stream or current through the Frozen Sea of such swiftness, as a Colmax told him, that if you cast anything therein, it would presently be carried out of sight towards the west.

Whereunto I answered, that there doth the like from Palus Maeotis, by the Euxine, the Bosphorus, and along the coast of Greece, etc., as it is affirmed by Contarenus, and divers others that have had experience of the same; and yet that sea lieth not open to any main sea that way, but is maintained by freshets, as by the Don, the Danube, etc.

In like manner is this current in the Frozen Sea increased and maintained by the Dwina, the river Ob, etc.

Now as I have here briefly recited the reasons alleged to prove a passage to Cathay by the north-east with my several answers thereunto, so will I leave it unto your judgment, to hope or despair of either at your pleasure.

CHAPTER IX.
HOW THAT THE PASSAGE BY THE NORTH-WEST IS MORE COMMODIOUS FOR OUR TRAFFIC THAN THE OTHER BY THE EAST, IF THERE WERE ANY SUCH.

1. By the north-east, if your winds do not give you a marvellous speedy and lucky passage, you are in danger (of being so near the Pole) to be benighted almost the one half of the year, and what danger that were, to live so long comfortless, void of light (if the cold killed you not), each man of reason or understanding may judge.

2. Also Mangia, Quinzai, and the Moluccas, are nearer unto us by the north-west than by the north-east more than two-fifths, which is almost by the half.

3. Also we may have by the rest a yearly return, it being at all times navigable, whereas you have but four months in the whole year to go by the north-east, the passage being at such elevation as it is formerly expressed, for it cannot be any nearer the south.

4. Furthermore, it cannot be finished without divers winterings by the way, having no havens in any temperate climate to harbour in there, for it is as much as we can well sail from hence to S. Nicholas, in the trade of Muscovy, and return in the navigable season of the year, and from S. Nicholas, Ciremissi, Tartarii, which standeth 80 degrees of the septentrional latitude, it is at the left 400 leagues, which amounteth scarce to the third part of the way, to the end of your voyage by the north-east.

5. And yet, after you have doubled this Cape, if then there might be found a navigable sea to carry you south-east according to your desire, yet can you not winter conveniently until you come to sixty degrees and to take up one degree running south-east you must sail twenty-four leagues and three four parts, which amounteth to four hundred and ninety-five leagues.

6. Furthermore, you may by the north-west sail thither, with all easterly winds, and return with any westerly winds, whereas you must have by the north-east sundry winds, and those proper, according to the lie of the coast and capes, you shall be enforced to double, which winds are not always to be had when they are looked for; whereby your journey should be greatly prolonged, and hardly endured so near the Pole, as we are taught by Sir Hugh Willoughbie, who was frozen to death far nearer the south.

7. Moreover, it is very doubtful whether we should long enjoy that trade by the north-east if there were any such passage that way, the commodities thereof once known to the Muscovite, what privilege soever he hath granted, seeing pollice with the maze of excessive gain, to the enriching of himself and all his dominions, would persuade him to presume the same, having so great opportunity, to distribute the commodities of those countries by the Naruc.

But by the north-west we may safely trade without danger or annoyance of any prince living, Christian or heathen, it being out of all their trades.

8. Also the Queen's Majesty's dominions are nearer the North-West Passage than any other great princes that might pass that way, and both in their going and return they must of necessity succour themselves and their ships upon some part of the same if any tempestuous weather should happen.

Further, no prince's navy of the world is able to encounter the Queen's Majesty's navy as it is at this present; and yet it should be greatly increased by the traffic ensuing upon this discovery, for it is the long voyages that increase and maintain great shipping.

Now it seemeth unnecessary to declare what commodities would grow thereby if all these things were as we have heretofore presupposed and thought them to be; which next adjoining are briefly declared.

CHAPTER X.
WHAT COMMODITIES WOULD ENSUE, THIS PASSAGE ONCE DISCOVERED.

1. It were the only way for our princes to possess the wealth of all the east parts (as they term them) of the world, which is infinite; as appeareth by the experience of Alexander the Great in the time of his conquest of India and the east parts of the world, alleged by Quintus Curtius, which would be a

great advancement to our country, wonderful enriching to our prince, and unspeakable commodities to all the inhabitants of Europe.

2. For, through the shortness of the voyage, we should be able to sell all manner of merchandise brought from thence far better cheap than either the Portuguese or Spaniard doth or may do. And, further, share with the Portuguese in the east and the Spaniard in the west by trading to any part of America through Mare del Sur, where they can no manner of way offend us.

3. Also we sailed to divers marvellous rich countries, both civil and others, out of both their jurisdictions, trades and traffics, where there is to be found great abundance of gold, silver, precious stones, cloth of gold, silks, all manner of spices, grocery wares, and other kinds of merchandise of an inestimable price, which both the Spaniard and Portuguese, through the length of their journeys, cannot well attain unto.

4. Also, we might inhabit some part of those countries, and settle there such needy people of our country which now trouble the commonwealth, and through want here at home are enforced to commit outrageous offences, whereby they are daily consumed with the gallows.

5. Moreover, we might from all the aforesaid places have a yearly return, inhabiting for our staple some convenient place of America, about Sierra Nevada or some other part, whereas it shall seem best for the shortening of the voyage.

6. Beside the exporting of our country commodities, which the Indians, etc., much esteem, as appeareth in Esther, where the pomp is expressed of the great King of India, Ahasuerus, who matched the coloured clothes wherewith his houses and tents were apparelled with gold and silver, as part of his greatest treasure, not mentioning velvets, silks, cloth of gold, cloth of silver, or such like, being in those countries most plentiful, whereby it plainly appeareth in what great estimation they would have the cloths of this our country, so that there would be found a far better vent for them by this means than yet this realm ever had; and that without depending either upon France, Spain, Flanders, Portugal, Hamborough, Emden, or any other part of Europe.

7. Also here we shall increase both our ships and mariners without burdening of the State.

8. And also have occasion to set poor men's children to learn handicrafts, and thereby to make trifles and such like, which the Indians and those people do much esteem; by reason whereof, there should be none occasion to have our country cumbered with loiterers, vagabonds, and such like idle persons.

All these commodities would grew by following this our discovery without injury done to any Christian prince by crossing them in any of their used trades, whereby they might take any just occasion of offence.

Thus have I briefly showed you some part of the grounds of my opinion, trusting that you will no longer judge me fantastic in this matter, seeing I have conceived no hope of this voyage, but am persuaded thereunto by the best cosmographers of our age, the same being confirmed both by reason and certain experiences.

Also this discovery hath been divers times heretofore by others both proposed, attempted, and performed.

It hath been proposed by Stephen Gomez unto Carolus, the fifth emperor in the year of our Lord 1527, as Alphonse Ullva testifieth in the story of Carolus' life, who would have set him forth in it (as the story mentioneth) if the great want of money, by reason of his long wars, had not caused him to surcease the same.

And the King of Portugal, fearing lest the emperor would have persevered in this his enterprise, gave him, to leave the matter unattempted, the sum of 350,000 crowns; and it is to be supposed that the King of Portugal would not have given to the emperor such sums of money for eggs in moonshine.

It hath been attempted by Corterialis the Portuguese, Scolmus the Dane, and by Sebastian Cabot in the time of King Henry VII.

And it hath been performed by the three brethren, the Indians aforesaid, and by Urdaneta, the friar of Mexico.

Also divers have proposed the like unto the French king, who hath sent two or three times to have discovered the same; the discoverers spending and consuming their victuals in searching the gulfs and bays between Florida and Labrador, whereby the ice is broken to the after-comers.

So that the right way may now be easily found out in short time, and that with little jeopardy and less expenses.

For America is discovered so far towards the north as Cape Frido, which is at 62 degrees, and that part of Greenland next adjoining is known to stand but at 72 degrees; so that we have but 10 degrees to sail north and south to put the world out of doubt hereof; and it is likely that the King of Spain and the King of Portugal would not have sat out all this while but that they are sure to possess to themselves all that trade they now use, and fear to deal in this discovery lest the Queen's Majesty, having so good opportunity, and finding the commodity which thereby might ensue to the commonwealth, would cut them off and enjoy the whole traffic to herself, and thereby the Spaniards and Portuguese with their great charges should beat the bush and other men catch the birds; which thing they foreseeing, have commanded that no pilot of theirs, upon pain of death, should seek to discover to the north-west, or plat out in any sea-card any through passage that way by the north-west.

Now, if you will impartially compare the hope that remaineth to animate me to this enterprise with those likelihoods which Columbus alleged before Ferdinando, the King of Castilia, to prove that there were such islands in the West Ocean as were after by him and others discovered, to the great commodity of Spain and all the world, you will think then that this North-West Passage to be most worthy travel therein.

For Columbus had none of the West Islands set forth unto him either in globe or card, neither yet once mentioned of any writer (Plato excepted, and the commentaries upon the same) from 942 years before Christ until that day.

Moreover, Columbus himself had neither seen America nor any other of the islands about it, neither understood he of them by the report of any other that had seen them, but only comforted himself with this hope, that the land had a beginning where the sea had an ending. For as touching that which the Spaniards do write of a Biscaine which should have taught him the way thither, it is thought to be imagined of them to deprive Columbus of his honour, being none of their countryman, but a stranger born.

And if it were true of the Biscaine, yet did he but hit upon the matter, or, at the least, gathered the knowledge of it by conjectures only.

And albeit myself have not seen this passage, or any part thereof, but am ignorant of it as touching experience as Columbus was before his attempt was made, yet have I both the report, relation, and authority of divers most credible men, which have both seen and passed through some and every part of this discovery, besides sundry reasons for my assurance thereof, all which Columbus wanted.

These things considered and impartially weighed together, with the wonderful commodities which this discovery may bring, especially to this realm of England, I must needs conclude with learned Baptista Ramusius, and divers other learned men, who said that this discovery hath been reserved for some noble prince or worthy man, thereby to make himself rich, and the world happy: desiring you to accept in good part this brief and simple discourse, written in haste, which, if I may perceive that it shall not sufficiently satisfy you in this behalf, I will then impart unto you a large discourse, which I have written only of this discovery.

And further, because it sufficeth not only to knew that such a thing there is, without ability to perform the same, I will at leisure make you partaker of another simple discourse of navigation, wherein I have not a little travelled, to make myself as sufficient to bring these things to effect as I have been ready to offer myself therein.

And therein I have devised to amend the errors of usual sea-cards, whose common fault is to make the degrees of longitude in every latitude of one like bigness.

And have also devised therein a spherical instrument, with a compass of variation for the perfect knowing of the longitude.

And a precise order to prick the sea-card, together with certain infallible rules for the shortening of any discovery, to know at the first entering of any strait whether it lies open to the ocean more ways than one, how far soever the sea stretcheth itself into the land.

Desiring you hereafter never to mislike with me for the taking in hand of any laudable and honest enterprise, for if, through pleasure and idleness, we purchase shame, the pleasure vanisheth, but the shame remaineth for ever.

And therefore, to give me leave without offence always to live and die in this mind, *that he is not worthy to live at all that for fear or danger of death shunneth his country's service and his own honour*, seeing death is inevitable, and the fame of virtue immortal. Wherefore, in this behalf, *Mutare vel timere sperno*.

CERTAIN OTHER REASONS OR ARGUMENTS TO PROVE A PASSAGE BY THE NORTH-WEST.

Learnedly written by Master Richard Willes, Gentleman.

Four famous ways there be spoken of to those fruitful and wealthy islands, which we do usually call Moluccas, continually haunted for gain, and daily travelled for riches therein growing. These islands, although they stand east from the meridian, distant almost half the length of the world, in extreme heat under the equinoctial line, possessed of infidels and barbarians, yet by our neighbours great abundance of wealth there is painfully sought in respect of the voyage dearly bought, and from thence dangerously brought home to us. Our neighbours I call the Portuguese, in comparison of the Molucchians for nearness unto us, for like situation westward as we have for their usual trade with us; for that the far south-easterings do know this part of Europe by no other name than Portugal, not greatly acquainted as yet with the other nations thereof. Their voyage is very well understood of all men, and the south-eastern way round about Africa, by the Cape of Good Hope, more spoken of, better known and travelled, than that it may seem needful to discourse thereof any farther.

The second way lieth south-west, between the West Indies, or South America, and the south continent, through that narrow strait where Magellan, first of all men that ever we do read of, passed these latter years, caving thereunto therefore his name. This way, no doubt, the Spaniards would commodiously take, for that it lieth near unto their dominions there, could the eastern current and Levant winds as easily suffer men to return as speedily therewith they may be carried thither; for the which difficulty, or rather impossibility of striving against the force both of wind and stream, this passage is little or nothing used, although it be very well known.

The third way, by the north-east, beyond all Europe and Asia, that worthy and renowned knight Sir Hugh Willoughbie sought to his peril, enforced there to end his life for cold, congealed and frozen to death. And, truly, this way consisteth rather in the imagination of geographers than allowable either in reason, or approved by experience, as well it may appear by the dangerous trending of the Scythian Cape set by Ortellius under the 80th degree north, by the unlikely sailing in that northern sea, always clad with ice and snow, or at the least continually pestered therewith, if haply it be at any time dissolved, beside bays and shelves, the water waxing more shallow towards the east, to say nothing of the foul mists and dark fogs in the cold clime, of the little power of the sun to clear the air, of the uncomfortable nights, so near the Pole, five months long.

A fourth way to go unto these aforesaid happy islands, the Moluccas, Sir Humphrey Gilbert, a learned and valiant knight, discourseth of at large in his new "Passage to Cathay." The enterprise of itself being virtuous, the fact must doubtless deserve high praise, and whensoever it shall be finished the fruits thereof cannot be small; where virtue is guide, there is fame a follower, and fortune a companion. But the way is dangerous, the passage doubtful, the voyage not thoroughly known, and therefore gainsaid by many, after this manner.

First, who can assure us of any passage rather by the north-west than by the north-east? do not both ways lie in equal distance from the North Pole? stand not the North Capes of either continent under like elevation? is not the ocean sea beyond America farther distant from our meridian by thirty or forty degrees west than the extreme points of Cathay eastward, if Ortellius' general card of the world be true? In the north-east that noble knight—Sir Hugh Willoughbie perished for cold, and can you then promise a passenger any better hap by the north-west, who hath gone for trial's sake, at any time, this way out of Europe to Cathay?

If you seek the advice herein of such as make profession in cosmography, Ptolemy, the father of geography, and his eldest children, will answer by their maps with a negative, concluding most of the sea within the land, and making an end of the world northward, near the 63rd degree. The same opinion, when learning chiefly flourished, was received in the Romans' time, as by their poets' writings it may appear. "Et te colet ultima Thule," said Virgil, being of opinion that Iceland was the extreme part of the world habitable toward the north. Joseph Moletius, an Italian, and Mercator, a German, for knowledge men able to be compared with the best geographers of our time, the one in his half spheres of the whole world, the other in some of his great globes, have continued the West Indies land, even to the North Pole, and consequently cut off all passage by sea that way.

The same doctors, Mercator in other of his globes and maps, Moletius in his sea-card, nevertheless doubting of so great continuance of the former continent, have opened a gulf betwixt the West Indies and the extreme northern land; but such a one that either is not to be travelled for the causes in the first objection alleged, or clean shut up from us in Europe by Greenland, the south end whereof Moletius maketh firm land with America, the north part continent with Lapland and Norway.

Thirdly, the greatest favourers of this voyage cannot deny but that, if any such passage be, it lieth subject unto ice and snow for the most part of the year, whereas it standeth in the edge of the frosty zone. Before the sun hath warmed the air and dissolved the ice, each one well knoweth that there can be no sailing; the ice once broken through the continual abode, the sun maketh a certain season in those parts. How shall it be possible for so weak a vessel as a ship is to hold out amid whole islands, as it were, of ice continually beating on each side, and at the mouth of that gulf, issuing down furiously from the north, safely to pass, when whole mountains of ice and snow shall be tumbled down upon her?

Well, grant the West Indies not to continue continent unto the Pole, grant there be a passage between these two lands, let the gulf lie nearer us than commonly in cards we find it set, namely, between the sixty-first and sixty-fourth degrees north, as Gemma Frisius in his maps and globes imagineth it, and so left by our countryman Sebastian Cabot in his table which the Earl of Bedford hath at Theinies; let the way be void of all difficulties, yet doth it not follow that we have free passage to Cathay. For example's sake, you may coast all Norway, Finmarke, and Lapland, and then bow southward to St. Nicholas, in Moscovy. You may likewise in the Mediterranean Sea fetch Constantinople and the mouth of the Don, yet is there no passage by sea through Moscovy into Pont Euxine, now called Mare Maggiore. Again, in the aforesaid Mediterranean Sea we sail to Alexandria in Egypt, the barbarians bring their pearl and spices from the Moluccas up the Red Sea and Arabian Gulf to Suez, scarcely three days' journey from the aforesaid haven; yet have we no way by sea from Alexandria to the Moluccas for that isthmus or little trait of land between the two seas. In like manner, although the northern passage be free at sixty-one degrees latitude, and the west ocean beyond America, usually called Mare del Sur, known to be open at forty degrees elevation for the island of Japan, yea, three hundred leagues northerly of Japan, yet may there be land to hinder the through passage that way by sea, as in the examples aforesaid it falleth out, Asia and America there being joined together in one continent. Nor can this opinion seem altogether frivolous unto any one that diligently peruseth our cosmographers' doings. Josephus Moletius is of that mind, not only in his plain hemispheres of the world, but also in his sea-card. The French geographers in like manner be of the same opinion, as by their map cut out in form of a heart you may perceive as though the West Indies were part of Asia, which sentence well agreeth with that old conclusion in the schools, *Quid-quid præter Africum et Europam est, Asia est*, "Whatsoever land doth neither appertain unto Africa nor to Europe is part of Asia."

Furthermore, it were to small purpose to make so long, so painful, so doubtful a voyage by such a new found way, if in Cathay you should neither be suffered to land for silks and silver, nor able to fetch the Molucca spices and pearl for piracy in those seas. Of a law denying all aliens to enter into China, and forbidding all the inhabiters under a great penalty to let in any stranger into those countries, shall you read in the report of Galeotto Petera, there imprisoned with other Portuguese, as also in the Japanese letters, how for that cause the worthy traveller Xavierus bargained with a barbarian merchant for a great sum of pepper to be brought into Canton, a port in Cathay. The great and dangerous piracy used in those seas no man can be ignorant of that listeth to read the Japanese and Indian history.

Finally, all this great labour would be lost, all these charges spent in vain, if in the end our travellers might not be able to return again, and bring safely home into their own native country that wealth and riches they in foreign regions with adventure of goods and danger of their lives have sought for. By the north-east there is no way; the South-East Passage the Portuguese do hold, as the lords of those seas. At the south-west, Magellan's experience hath partly taught us, and partly we are persuaded by reason, how the eastern current striketh so furiously on that strait, and falleth with such force into that narrow gulf, that hardly any ship can return that way into our west ocean out of Mare del Sur. The which, if it be true, as truly it is, then we may say that the aforesaid eastern current, or Levant course of waters, continually following after the heavenly motions, loseth not altogether its force, but is doubled rather by another current from out the north-east, in the passage between America and the North Land, whither it is of necessity carried, having none other way to maintain itself in circular motion, and consequently the force and fury thereof to be no less in the Strait of Anian, where it striketh south into Mare del Sur beyond America (if any such strait of sea there be), than in the strait of Magellan, both straits being of like breadth, as in Belognine Salterius' table of "New France," and in Don Diego Hermano de Toledo's card for navigation in that region, we do find precisely set down.

Nevertheless, to approve that there lieth a way to Cathay at the north-west from out of Europe, we have experience, namely of three brethren that went that journey, as Gemma Frisius recordeth, and left a name unto that strait, whereby now it is called Fretum Trium Fratrum. We do read again of a Portuguese that passed this strait, of whom Master Frobisher speaketh, that was imprisoned therefore many years in Lisbon, to verify the old Spanish proverb, "I suffer for doing well." Likewise, An. Urdaneta, a friar of Mexico, came out of Mare del Sur this way into Germany; his card, for he was a great discoverer, made by his own experience and travel in that voyage, hath been seen by gentlemen of good credit.

Now if the observation and remembrance of things breedeth experience, and of experience proceedeth art, and the certain knowledge we have in all faculties, as the best philosophers that ever were do affirm truly the voyage of these aforesaid travellers that have gone out of Europe into Mare del Sur, and returned thence at the north-west, do most evidently conclude that way to be navigable, and that passage free; so much the more we are so to think, for that the first principle and chief ground in all geography, as Ptolemy saith, is the history of travel, that is, reports made by travellers skilful in geography and astronomy, of all such things in their journey as to geography do belong. It only remaineth, that we now answer to those arguments that seemed to make against this former conclusion.

The first objection is of no force, that general table of the world, set forth by Ortellius or Mercator, for it greatly skilleth not, being unskilfully drawn for that point, as manifestly it may appear unto any one that compareth the same with Gemma Frisius' universal map, with his round quartered card, with his globe, with Sebastian Cabot's table, and Ortellius' general map alone, worthily preferred in this case before all Mercator's and Ortellius' other doings: for that Cabot was not only a skilful seaman, but a long traveller, and such a one as entered personally that strait, sent by King Henry VII. to make this aforesaid discovery, as in his own discourse of navigation you may read in his card drawn with his own hand, that the mouth of the north-western strait lieth near the 318th meridian, between 61 and 64 degrees in the elevation, continuing the same breadth about ten degrees west, where it openeth southerly more and more, until it come under the tropic of Cancer; and so runneth into Mare del Sur, at the least 18 degrees more in breadth there than it was where it first began; otherwise I could as well imagine this passage to be more unlikely than the voyage to Moscovy, and more impossible than it for the far situation and continuance thereof in the frosty clime: as now I can affirm it to be very possible and most likely in comparison thereof, for that it neither coasteth so far north as the Moscovian passage doth, neither is this strait so long as that, before it bow down southerly towards the sun again.

The second argument concludeth nothing. Ptolemy knew not what was above 16 degrees south beyond the equinoctial line, he was ignorant of all passages northward from the elevation of 63 degrees, he knew no ocean sea beyond Asia, yet have the Portuguese trended the Cape of Good Hope at the south point of Africa, and travelled to Japan, an island in the east ocean, between Asia and America; our merchants in the time of King Edward the Sixth discovered the Moscovian passage farther north than Thule, and showed Greenland not to be continent with Lapland and Norway: the like our north-western travellers have done, declaring by their navigation that way the ignorance of all cosmographers that either do join Greenland with America, or continue the West Indies with that frosty region under the North Pole. As for Virgil, he sang according to the knowledge of men in his time, as another poet did of the hot zone.

Quarum quæ media est, non est habitabilis æstu. Imagining, as most men then did, Zonam Torridam, the hot zone, to be altogether dishabited for heat, though presently we know many famous and worthy kingdoms and cities in that part of the earth, and the island of S. Thomas near Ethiopia, and the wealthy islands for the which chiefly all these voyages are taken in hand, to be inhabited even under the equinoctial line.

To answer the third objection, besides Cabot and all other travellers' navigations, the only credit of Master Frobisher may suffice, who lately, through all these islands of ice and mountains of snow, passed that way, even beyond the gulf that tumbleth down from the north, and in some places, though he drew one inch thick ice, as he returning in August did, came home safely again.

The fourth argument is altogether frivolous and vain, for neither is there any isthmus or strait of land between America and Asia, nor can these two lands jointly be one continent. The first part of my answer is manifestly allowed by Homer, whom that excellent geographer, Strabo, followeth, yielding him in this faculty the prize. The author of that book likewise *On the Universe* to Alexander, attributed unto Aristotle, is of the same opinion that Homer and Strabo be of, in two or three places. Dionysius, in his *Periegesis*, hath this verse, "So doeth the ocean sea run round about the world:" speaking only of Europe, Africa, and Asia, as then Asia was travelled and known. With these doctors may you join Pomponius Mela, Pliny, Pius, in his description of Asia. All the which writers do no less confirm the whole eastern side of Asia to be compassed about with the sea; then Plato doth affirm in is *Timaeus*, under the name Atlantis, the West Indies to be an island, as in a special discourse thereof R. Eden writeth, agreeable unto the sentence of Proclus, Marsilius Ficinus, and others. Out of Plato it is gathered that America is an island. Homer, Strabo, Aristotle, Dionysius, Mela, Pliny, Pius, affirm the continent of Asia, Africa, and Europe, to be environed with the ocean. I may therefore boldly say (though later intelligences thereof had we none at all) that Asia and the West Indies be not tied together by any isthmus or strait of land, contrary to the opinion of some new cosmographers, by whom doubtfully this matter hath been brought in controversy. And thus much for the first part of my answer unto the fourth objection.

The second part, namely, that America and Asia cannot be one continent, may thus be proved:—"The most rivers take down that way their course, where the earth is most hollow and deep," writeth Aristotle; and the sea (saith he in the same place), as it goeth further, so is it found deeper. Into what gulf do the Moscovian rivers Onega, Dwina, Ob, pour out their streams? northward out of Moscovy into the sea. Which way doth that sea strike? The south is main land, the eastern coast waxeth more and more shallow: from the north, either naturally, because that part of the earth is higher, or of necessity, for that the forcible influence of some northern stars causeth the earth there to shake off the sea, as some philosophers do think; or, finally, for the great store of waters engendered in that frosty and cold climate, that the banks are not able to hold them. From the north, I say, continually falleth down great abundance of water; so this north-eastern current must at the length abruptly bow toward us south on the west side of Finmark and Norway, or else strike down south-west above Greenland, or betwixt Greenland and Iceland, into the north-west strait we speak of, as of congruence it doth, if you mark the situation of that region, and by the report of Master Frobisher experience teacheth us. And, Master Frobisher, the further he travelled in the former passage, as he told me, the deeper always he found the sea. Lay you now the sum hereof together, the rivers run where the channels are most hollow, the sea in taking his course waxeth deeper, the sea waters fall continually from the north southward, the north-eastern current striketh down into the strait we speak of and is there augmented with whole mountains of ice and snow falling down furiously out from the land under the North Pole. Where store of water is, there is it a thing impossible to want sea; where sea not only doth not want, but waxeth deeper, there can be discovered no land. Finally, whence I pray you came the contrary tide, that Master Frobisher met withal, after that he had sailed no small way in that passage, if there be any isthmus or strait of land betwixt the aforesaid north-western gulf and Mare del Sur, to join Asia and America together? That conclusion arrived at in the schools, "Whatsoever land doth neither appertain unto Africa, nor to Europe, is part of Asia," was meant of the parts of the world then known, and so is it of right to be understood.

The fifth objection requireth for answer wisdom and policy in the traveller to win the barbarians' favour by some good means; and so to arm and strengthen himself, that when he shall have the repulse in one coast, he may safely travel to another, commodiously taking his convenient times, and discreetly making choice of them with whom he will thoroughly deal. To force a violent entry would for us Englishmen be very hard, considering the strength and valour of so great a nation, far distant from us, and the attempt thereof might be most perilous unto the doers, unless their park were very good.

Touching their laws against strangers, you shall read nevertheless in the same relations of Galeotto Perera, that the Cathaian king is wont to grant free access unto all foreigners that trade into his country for merchandise, and a place of liberty for them to remain in; as the Moors had, until such time as they had brought the Loutea or Lieutenant of that coast to be a circumcised Saracen: wherefore some of them were put to the sword, the rest were scattered abroad; at Fuquien, a great city in China, certain of them are yet this day to be seen. As for the Japanese, they be most desirous to be acquainted with strangers. The Portuguese, though they were straitly handled there at the first, yet in the end they found great favour at the prince's hands, insomuch that the Loutea or President that misused them was therefore put to death. The rude Indian canoe voyageth in those seas, the Portuguese, the

Saracens, and Moors travel continually up and down that reach from Japan to China, from China to Malacca, from Malacca to the Moluccas, and shall an Englishman better appointed than any of them all (that I say no more of our navy) fear to sail in that ocean? what seat at all do want piracy? what navigation is there void of peril?

To the last argument our travellers need not to seek their return by the north-east, neither shall they be constrained, except they list, either to attempt Magellan's strait at the south-west, or to be in danger of the Portuguese on the south-east; they may return by the north-west, that same way they do go forth, as experience hath showed.

The reason alleged for proof of the contrary may be disposed after this manner: And first, it may be called in controversy, whether any current continually be forced by the motion of primum mobile, round about the world or no; for learned men do diversely handle that question. The natural course of all waters is downward, wherefore of congruence they fall that way where they find the earth most low and deep: in respect whereof, it was erst said, the seas do strike from the northern lands southerly. Violently the seas are tossed and troubled divers ways with the winds, increased and diminished by the course of the moon, hoisted up and down through the sundry operations of the sun and the stars: finally, some be of opinion that the seas be carried in part violently about the world, after the daily motion of the highest movable heaven, in like manner as the elements of air and fire, with the rest of the heavenly spheres, are from the east unto the west. And this they do call their eastern current, or Levant stream. Some such current may not be denied to be of great force in the hot zone, for the nearness thereof unto the centre of the sun, and blustering eastern winds violently driving the seas westward; howbeit in the temperate climes the sun being farther off, and the winds more diverse, blowing as much from the north, the west, and south, as from the east, this rule doth not effectually withhold us from travelling eastwards, neither be we kept ever back by the aforesaid Levant winds and stream. But in Magellan strait we are violently driven back westward, ergo through the north-western strait or Anian frith shall we not be able to return eastward: it followeth not. The first, for that the north-western strait hath more sea room at the least by one hundred English miles than Magellan's strait hath, the only want whereof causeth all narrow passages generally to be most violent. So would I say in the Anian Gulf, if it were so narrow as Don Diego and Zalterius have painted it out, any return that way to be full of difficulties, in respect of such straitness thereof, not for the nearness of the sun or eastern winds, violently forcing that way any Levant stream; but in that place there is more sea room by many degrees, if the cards of Cabot and Gemma Frisius, and that which Tramezine imprinted, be true.

And hitherto reasons see I none at all, but that I may as well give credit unto their doings as to any of the rest. It must be *Peregrinationis historia*, that is, true reports of skilful travellers, as Ptolemy writeth, that in such controversies of geography must put us out of doubt. Ortellius, in his universal tables, in his particular maps of the West Indies, of all Asia, of the northern kingdoms, of the East Indies; Mercator in some of his globes and general maps of the world, Moletius in his universal table of the Globe divided, in his sea-card and particular tables of the East Indies Zanterius and Don Diego with Fernando Bertely, and others, do so much differ both from Gemma Frisius and Cabot among themselves, and in divers places from themselves, concerning the divers situation and sundry limits of America, that one may not so rashly as truly surmise these men either to be ignorant in those points touching the aforesaid region, or that the maps they have given out unto the world were collected only by them, and never of their own drawing.

THE FIRST VOYAGE OF MASTER MARTIN FROBISHER

To the North-West for the search of the passage or strait to China, written by Christopher Hall, and made in the year of our Lord 1576.

Upon Monday, the thirteenth of May, the barque *Gabriel* was launched at Redriffe, and upon the twenty-seventh day following she sailed from Redriffe to Ratcliffe.

The seventh of June being Thursday, the two barques, viz., the *Gabriel* and the *Michael*, and our pinnace, set sail at Ratcliffe, and bare down to Deptford, and there we anchored. The cause was, that our pinnace burst her bowsprit and foremast aboard of a ship that rowed at Deptford, else we meant to have passed that day by the court, then at Greenwich.

The eighth day being Friday, about twelve o'clock, we weighed at Deptford and set sail all three of us and bare down by the court, where we shot off our ordinance, and made the best show we could; her Majesty beholding the same commended it, and bade us farewell with shaking her hand at us out of the window. Afterwards she sent a gentleman aboard of us, who declared that her Majesty had good liking of our doings, and thanked us for it, and also willed our captain to come the next day to the court to take his leave of her.

The same day, towards night, Master Secretary Woolley came aboard of us, and declared to the company that her Majesty had appointed him to give them charge to be obedient, and diligent to their captain and governors in all things, and wished us happy success.

The ninth day about noon, the wind being westerly, having our anchors aboard ready to set sail to depart, we wanted some of our company, and therefore stayed and moored them again.

Sunday, the tenth of June, we set sail from Blackwall at a south-west and by west sun, the wind being at north-north-west, and sailed to Gravesend, and anchored there at a west-north-west sun, the wind being as before.

The twelfth day, being over against Gravesend, by the Castle or Blockhouse, we observed the latitude, which was 51 degrees 33 minutes, and in that place the variation of the compass is 11 degrees and a half. This day we departed from Gravesend at a west-south-west sun, the wind at north and by east a fair gale, and sailed to the west part of Tilbury Hope, and so turned down the Hope, and at a west sun the wind came to the east-south-east, and we anchored in seven fathoms, being low water.

[Here there follows an abstract of the ship's log, showing the navigation until the 28th of July, when they had sight of land supposed to be Labrador.]

July 28th. From 4 to 8, 4 leagues: from 8. to 12, 3 leagues: from 12 to 4, north and by west, 6 leagues, but very foggy; from thence to 8 of the clock in the morning little wind, but at the clearing up of the fog we had sight of land, which I supposed to be Labrador, with great store of ice about the land; I ran in towards it, and sounded, but could get no land at 100 fathoms, and the ice being so thick I could not get to the shore, and so lay off and came clear of the ice. Upon Monday we came within a mile of the shore, and sought a harbour; all the sound was full of ice, and our boat rowing ashore could get no ground at 100 fathom, within a cable's length of the shore; then we sailed east-north-east along the shore, for so the land lieth, and the current is there great, setting north-east and south-west; and if we could have gotten anchor ground we would have seen with what force it had run, but I judge a ship may drive a league and a half in one hour with that tide.

This day, at four of the clock in the morning, being fair and clear, we had sight of a headland as we judged bearing from us north and by east, and we sailed north-east and by north to that land, and when we came thither we could not get to the land for ice, for the ice stretched along the coast, so that we could not come to the land by 5 leagues.

Wednesday, the first of August, it calmed, and in the afternoon I caused my boat to be hoisted out, being hard by a great island of ice, and I and four men rowed to that ice, and sounded within two cables' length of it, and had 16 fathoms and little stones, and after that sounded again within a minion's shot, and had ground at 100 fathoms, and fair sand. We sounded the next day a quarter of a mile from it, and had 60 fathoms rough ground, and at that present being aboard, that great island of ice fell one part from another, making a noise as if a great cliff had fallen into the sea. And at 4 of the clock I sounded again, and had 90 fathoms, and small black stones, and little white stones like pearls. The tide here did set to the shore.

We sailed this day south-south-east ofward, and laid it a tric.

The next day was calm and thick, with a great sea.

The next day we sailed south and by east two leagues, and at 8 of the clock in the forenoon we cast about to the eastward.

The sixth day it cleared, and we ran north-west into the shore to get a harbour, and being towards night, we notwithstanding kept at sea.

The seventh day we plied room with the shore, but being near it it waxed thick, and we bare off again.

The eighth day we bended in towards the shore again.

The ninth day we sounded, but could get no ground at 130 fathoms. The weather was calm.

The tenth I took four men and myself, and rode to shore, to an island one league from the main, and there the flood setteth south-west along the shore, and it floweth as near as I could judge so too. I could not tarry to prove it, because the ship was a great way from me, and I feared a fog; but when I came ashore it was low water. I went to the top of the islands and before I came back it was hied a foot water, and so without tarrying I came aboard.

The eleventh we found our latitude to be 63 degrees and 8 minutes, and this day entered the strait.

The twelfth we set sail towards an island called the Gabriel's Island, which was 10 leagues then from us.

We espied a sound, and bare with it, and came to a sandy bay, where we came to an anchor, the land bearing east-south-east of us, and there we rode all night in 8 fathom water. It floweth there at a south-east moon; we called it Prior's Sound, being from the Gabriel's Island 10 leagues.

The fourteenth we weighed and ran into another sound, where we anchored in 8 fathoms water, fair sand, and black ooze, and there caulked our ship, being weak from the gunwales upward, and took in fresh water.

The fifteenth day we weighed, and sailed to Prior's Bay, being a mile from thence.

The sixteenth day was calm, and we rode still without ice, but presently within two hours it was frozen round about the ship, a quarter of an inch thick, and that bay very fair and calm.

The seventeenth day we weighed, and came to Thomas William's Island.

The eighteenth day we sailed north-north-west and anchored again in 23 fathoms, and caught ooze under Bircher's Island, which is from the former island 10 leagues.

The nineteenth day in the morning, being calm, and no wind, the captain and I took our boat, with eight men in her, to row us ashore, to see if there were there any people, or no, and going to the top of the island, we had sight of seven boats, which came rowing from the east side toward that island; whereupon we returned aboard again. At length we sent our boat, with five men in her, to see whither they rowed, and so with a white cloth brought one of their boats with their men along the shore, rowing after our boat, till such time as they saw our ship, and then they rowed ashore. Then I went on shore myself, and gave every of them a threaden point, and brought one of them aboard of me, where he did eat and drink, and then carried him on shore again. Whereupon all the rest came aboard with their boats, being nineteen persons, and they spake, but we understood them not. They be like to Tartars, with long black hair, broad faces, and flat noses, and tawny in colour, wearing seal skins, and so do the women, not differing in the fashion, but the women are marked in the face with blue streaks down the cheeks and round about the eyes. Their boats are made all of seal skins, with a keel of wood within the skin: the proportion of them is like a Spanish shallop, save only they be flat in the bottom and sharp at both ends.

The twentieth day we weighed, and went to the east side of this island, and I and the captain, with four men more, went on shore, and there we saw their houses, and the people espying us, came rowing towards our boat, whereupon we plied to our boat; and we being in our boat and they ashore, they called to us, and we rowed to them, and one of their company came into our boat, and we carried him aboard, and gave him a bell and a knife; so the captain and I willed five of our men to set him ashore at a rock, and not among the company which they came from, but their wilfulness was such that they would go to them, and so were taken themselves and our boat lost.

The next day in the morning we stood in near the shore and shot off a fauconet, and sounded our trumpet, but we could hear nothing of our men. This sound we called the Five Men's Sound, and plied out of it, but anchored again in 30 fathoms and ooze; and riding there all night, in the morning the snow lay a foot thick upon our hatches.

The two-and-twentieth day in the morning we weighed, and went again to the place where we lost our men and our boat. We had sight of fourteen boats, and some came near to us, but we could learn nothing of our men. Among the rest, we enticed one in a boat to our ship's side with a bell; and in giving him the bell we took him and his boat, and so kept him, and so rowed down to Thomas William's island, and there anchored all night.

The twenty-sixth day we weighed to come homeward, and by twelve of the clock at noon we were thwart of Trumpet's Island.

The next day we came thwart of Gabriel's Island, and at eight of the clock at night we had the Cape Labrador west from us ten leagues.

The twenty-eighth day we went our course south-east.

We sailed south-east and by east, twenty-two leagues.

The first day of September, in the morning, we had sight of the land of Friesland, being eight leagues from us, but we could not come nearer it for the monstrous ice that lay about it. From this day till the sixth of this month we ran along Iceland, and had the south part of it at eight of the clock east from us ten leagues.

The seventh day of this month we had a very terrible storm, by force whereof one of our men was blown into the sea out of our waste, but he caught hold of the foresail sheet, and there held till the captain plucked him again into the ship.

The twenty-fifth day of this month we had sight of the island of Orkney, which was then east from us.

The first day of October we had sight of the Sheld, and so sailed along the coast, and anchored at Yarmouth, and the next day we came into Harwich.

THE LANGUAGE OF THE PEOPLE OF META INCOGNITA.

Argotteyt, a hand.	Attegay, a coat.
Cangnawe, a nose.	Polleuetagay, a knife.
Arered, an eye.	Accaskay, a ship.
Keiotot, a tooth.	Coblone, a thumb.
Mutchatet, the head.	Teckkere, the foremost finger.
Chewat, an ear.	Ketteckle, the middle finger.
Comagaye, a leg.	Mekellacane, the fourth finger.
Atoniagay, a foot.	
Callagay, a pair of breeches.	Yachethronc, the little finger.

THE SECOND VOYAGE OF MASTER MARTIN FROBISHER,

Made to the West and North-West Regions in the year 1577, with a Description of the Country and People, written by Dionise Settle.

On Whit Sunday, being the sixth-and-twentieth day of May, in the year of our Lord God 1577, Captain Frobisher departed from Blackwall—with one of the Queen's Majesty's ships called the *Aid*, of nine score ton or thereabout, and two other little barques likewise, the one called the *Gabriel*, whereof Master Fenton, a gentleman of my Lord of Warwick's, was captain; and the other the *Michael*, whereof Master York, a gentleman of my lord admiral's, was captain, accompanied with seven score gentlemen, soldiers, and sailors, well furnished with victuals and other provisions necessary for one half year—on this, his second year, for the further discovering of the passage to Cathay and other countries thereunto adjacent, by west and north-west navigations, which passage or way is supposed to be on the north and north-west parts of America, and the said America to be an island environed with the sea, where through our merchants might have course and recourse with their merchandise from these our northernmost parts of Europe, to those Oriental coasts of Asia in much shorter time and with greater benefit than any others, to their no little commodity and profit that do or shall traffic the same. Our said captain and general of this present voyage and company, having the year before, with two little pinnaces to his great danger, and no small commendations, given a worthy attempt towards the performance thereof, is also pressed when occasion shall be ministered to the benefit of his prince and native country—to adventure himself further therein. As for this second voyage, it seemeth sufficient that he hath better explored and searched the commodities of those people and countries, with sufficient commodity unto the adventurers, which, in his first voyage the year before, he had found out.

Upon which considerations the day and year before expressed, he departed from Blackwall to Harwich, where making an accomplishment of things necessary, the last of May we hoisted up sails, and with a merry wind the 7th of June we arrived at the islands called Orchades, or vulgarly Orkney, being in number thirty, subject and adjacent to Scotland, where we made provision of fresh water, in the doing whereof our general licensed the gentlemen and soldiers, for their recreation, to go on shore. At our landing the people fled from their poor cottages with shrieks and alarms, to warn their neighbours of enemies, but by gentle persuasions we reclaimed them to their houses. It seemeth they are often frighted with pirates, or some other enemies, that move them to such sudden fear. Their houses are very simply built with pebble stone, without any chimneys, the fire being made in the midst thereof. The good man, wife, children, and other of their family, eat and sleep on the one side of the house, and their cattle on the other, very beastly and rudely in respect of civilisation. They are destitute of wood, their fire is turf and cow shardes. They have corn, bigge, and oats, with which they pay their king's rent to the maintenance of his house. They take great quantity of fish, which they dry in the wind and sun; they dress their meat very filthily, and eat it without salt. Their apparel is after the nudest sort of Scotland. Their money is all base. Their Church and religion is reformed according to the Scots. The fishermen of England can better declare the dispositions of those people than I, wherefore I remit other their usages to their reports, as yearly repairers thither in their courses to and from Iceland for fish.

We departed here hence the 8th of June, and followed our course between west and north-west until the 4th of July, all which time we had no night, but that easily, and without any impediment, we had, when we were so disposed, the fruition of our books, and other pleasures to pass away the time, a thing of no small moment to such as wander in unknown seas and long navigations, especially when both the winds and raging surges do pass their common and wonted course. This benefit endureth in those parts not six weeks, whilst the sun is near the tropic of Cancer, but where the pole is raised to 70 or 80 degrees it continueth the longer.

All along these seas, after we were six days sailing from Orkney, we met, floating in the sea, great fir trees, which, as we judged, were, with the fury of great floods, rooted up, and so driven into the sea. Iceland hath almost no other wood nor fuel but such as they take up upon their coasts. It seemeth that these trees are driven from some part of the Newfoundland, with the current that setteth from the west to the east.

The 4th of July we came within the making of Friesland. From this shore, ten or twelve leagues, we met great islands of ice of half a mile, some more, some less in compass, showing above the sea thirty or forty fathoms, and as we supposed fast on ground, where, with our lead, we could scarce sound the bottom for depth.

Here, in place of odoriferous and fragrant smells of sweet gums and pleasant notes of musical birds, which other countries in more temperate zones do yield, we tasted the most boisterous Boreal blasts, mixed with snow and hail, in the months of June and July, nothing inferior to our untemperate winter: a sudden alteration, and especially in a place of parallel, where the pole is not elevated above 61 degrees, at which height other countries more to the north, yea unto 70 degrees, show themselves more temperate than this doth. All along this coast ice lieth as a continual bulwark, and so defendeth the country, that those which would land there incur great danger. Our general, three days together, attempted with the ship boat to have gone on shore, which, for that without great danger he could not accomplish, he deferred it until a more convenient time. All along the coast lie very high mountains, covered with snow, except in such places where, through the steepness of the mountains, of force it must needs fall. Four days coasting along this land we found no sign of habitation. Little birds which we judged to have lost the shore, by reason of thick fogs which that country is much subject unto, came flying to our ships, which causeth us to suppose that the country is both more tolerable and also habitable within than the outward shore maketh show or signification.

From hence we departed the 8th of July, and the 16th of the same we came with the making of land, which land our general the year before had named the Queen's Forehand, being an island, as we judge, lying near the supposed continent with America, and on the other side, opposite to the same, one other island, called Halles Isle, after the name of the master of the ship, near adjacent to the firm land, supposed continent with Asia. Between the which two islands there is a large entrance or strait, called Frobisher's Strait, after the name of our general, the first finder thereof. This said strait is supposed to have passage into the sea of Sur, which I leave unknown as yet.

It seemeth that either here, or not far hence, the sea should have more large entrance than in other parts within the frozen or untemperate zone, and that some contrary tide, either from the east or west, with main force casteth out that great quantity of ice which cometh floating from this coast, even unto Friesland, causing that country to seem more untemperate than others much more northerly than the same.

I cannot judge that any temperature under the Pole, being the time of the Sun's northern declination, half a year together, and one whole day (considering that the sun's elevation surmounteth not twenty-three degrees and thirty minutes), can have power to dissolve such monstrous and huge ice, comparable to great mountains, except by some other force, as by swift currents and tides, with the help of the said day of half a year.

Before we came within the making of these lands, we tasted cold storms, insomuch that it seemed we had changed with winter, if the length of the days had not removed us from that opinion.

At our first coming, the straits seemed to be shut up with a long mure of ice, which gave no little cause of discomfort unto us all; but our general (to whose diligence, imminent dangers and difficult attempts seemed nothing in respect of his willing mind for the commodity of his prince and country), with two little pinnaces prepared of purpose, passed twice through them to the east shore, and the islands thereunto adjacent; and the ship, with the two barques, lay off and on something farther into the sea from the danger of the ice.

Whilst he was searching the country near the shore, some of the people of the country showed themselves, leaping and dancing, with strange shrieks and cries, which gave no little admiration to our men. Our general, desirous to allure them unto him by fair means, caused knives and other things to be proffered unto them, which they would not take at our hands; but being laid on the ground, and the party going away, they came and took up, leaving something of theirs to countervail the same. At the length, two them, leaving their weapons, came down to our general and master, who did the like to them, commanding the company to stay, and went unto them, who, after certain dumb signs and mute congratulations, began to lay hands upon them, but they deliverly escaped, and ran to their bows and arrows and came fiercely upon them, not respecting the rest of our company, which were ready for their defence, but with their arrows hurt divers of them. We took the one, and the other escaped.

Whilst our general was busied in searching the country, and those islands adjacent on the east shore, the ships and barques, having great care not to put far into the sea from him, for that he had small store of victuals, were forced to abide in a cruel tempest, chancing in the night amongst and in the thickest of the ice, which was so monstrous that even the least of a thousand had been of force sufficient to have shivered our ship and barques into small portions, if God (who in all necessities hath care upon the infirmity of man) had not provided for this our extremity a sufficient remedy, through the light of the night, whereby we might well discern to flee from such imminent dangers, which we avoided within fourteen bourdes in one watch, the space of four hours. If we had not incurred this danger amongst these monstrous islands of ice, we should have lost our general and master, and the most of our best sailors, which were on the shore destitute of victuals; but by the valour of our master gunner, Master Jackman and Andrew Dier, the master's mates, men expert both in navigation and other good qualities, we were all content to incur the dangers afore rehearsed, before we would, with our own safety, run into the seas, to the destruction of our said general and his company.

The day following, being the 19th of July, our captain returned to the ship with good news of great riches, which showed itself in the bowels of those barren mountains, wherewith we were all satisfied. A sudden mutation. The one part of us being almost swallowed up the night before, with cruel Neptune's force, and the rest on shore, taking thought for their greedy paunches how to find the way to Newfoundland; at one moment we were racked with joy, forgetting both where we were and what we had suffered. Behold the glory of man: to-night contemning riches, and rather looking for death than otherwise, and to-morrow devising how to satisfy his greedy appetite with gold.

Within four days after we had been at the entrance of the straits, the north-west and west winds dispersed the ice into the sea, and made us a large entrance into the Straits, that without impediment, on the 19th July, we entered them; and the 20th thereof our general and master, with great diligence, sought out and sounded the west shore, and found out a fair harbour for the ship and barques to ride in, and named it after our master's mate, Jackman's Sound, and brought the ship, barques, and all their company to safe anchor, except one man which died by God's visitation.

At our first arrival, after the ship rode at anchor, general, with such company as could well be spared from the ships, in marching order entered the land, having special care by exhortations that at our entrance thereinto we should all with one voice, kneeling upon our knees, chiefly thank God for our safe arrival; secondly, beseech Him that it would please His Divine Majesty long to continue our Queen, for whom he, and all the rest of our company, in this order took possession of the country; and thirdly, that by our Christian study and endeavour, those barbarous people, trained up in paganry and infidelity, might be reduced to the knowledge of true religion, and to the hope of salvation in Christ our Redeemer, with other words very apt to signify his willing mind and affection towards his prince and country, whereby all suspicion of an undutiful subject may credibly be judged to be utterly exempted from his mind. All the rest of the gentlemen, and others, deserve worthily herein their due praise and commendation.

These things in order accomplished, our general commanded all the company to be obedient in things needful for our own safeguard to Master Fenton, Master Yorke, and Master Beast, his lieutenant, while he was occupied in other necessary affairs concerning our coming thither.

After this order we marched through the country, with ensign displayed, so far as was thought needful, and now and then heaped up stones on high mountains and other places, in token of possession, as likewise to signify unto such as hereafter may chance to arrive there that possession is taken in the behalf of some other prince by those which first found out the country.

Whose maketh navigation to these countries hath not only extreme winds and furious seas to encounter withal, but also many monstrous and great islands of ice: a thing both rare, wonderful, and greatly to be regarded.

We were forced sundry times, while the ship did ride here at anchor, to have continual watch, with boats and men ready with hawsers, to knit fast unto such ice which with the ebb and flood were tossed to and fro in the harbour, and with force of oars to hail them away, for endangering the ship.

Our general certain days searched this supposed continent with America, and not finding the commodity to answer his expectations, after he had made trial thereof, he departed thence, with two little barques, and men sufficient, to the east shore, being he supposed continent of Asia, and left the ship, with most of the gentlemen soldiers and sailors, until such time as he either thought good to send or come for them.

The stones on this supposed continent with America be altogether sparkled and glister in the sun like gold; so likewise doth the sand in the bright water, yet they verify the old proverb, "All is not gold that glistereth."

On this west shore we found a dead fish floating, which had in his nose a horn, straight and torquet, of length two yards lacking two inches, being broken in the top, where we might perceive it hollow, into which some of our sailors putting spiders they presently died. I saw not the trial hereof, but it was reported unto me of a truth, by the virtue whereof we supposed it to be the sea unicorn.

After our general had found out good harbour for the ship and barques to anchor in, and also such store of gold ore as he thought himself satisfied withal, he returned to the *Michael*, whereof Master Yorke aforesaid was captain, accompanied with our master and his mate, who coasting along the west shore, not far from whence the ship rode, they perceived a fair harbour, and willing to sound the same, at the entrance thereof they espied two tents of seal skins, unto which the captain, our said master, and other company resorted. At the sight of our men the people fled into the mountains; nevertheless, they went to their tents, where, leaving certain trifles of ours as glasses, bells, knives, and such like things, they departed, not taking anything of theirs except one dog. They did in like manner leave behind them a letter, pen, ink, and paper, whereby our men whom the captain lost the year before, and in that people's custody, might (if any of them were alive) be advertised of our presence and being there.

On the same day, after consultation, all the gentlemen, and others likewise that could be spared from the ship, under the conduct and leading of Master Philpot (unto whom, in our general's absence, and his lieutenant, Master Beast, all the rest were obedient), went ashore, determining to see if by fair means we could either allure them to familiarity, or otherwise take some of them, and so attain to some knowledge of those men whom our general lost the year before.

At our coming back again to the place where their tents were before, they had removed their tents farther into the said bay or sound, where they might, if they were driven from the land, flee with their boats into the sea. We, parting ourselves into two companies, and compassing a mountain, came suddenly upon them by land, who, espying us, without any tarrying fled to their boats, leaving the most part of their oars behind them for haste, and rowed down the bay, where our two pinnaces met them and drove them to shore. But if they had had all their oars, so swift are they in rowing, it had been lost time to have chased them.

When they were landed they fiercely assaulted our men with their bows and arrows, who wounded three of them with our arrows, and perceiving themselves thus hurt they desperately leaped off the rocks into the sea and drowned themselves; which if they had not done but had submitted themselves, or if by any means we could have taken alive (being their enemies as they judged), we would both have saved them, and also have sought remedy to cure their wounds received at our hands. But they, altogether void of humanity, and ignorant what mercy meaneth, in extremities look for no other than death, and perceiving that they should fall into our hands, thus miserably by drowning rather desired death than otherwise to be saved by us. The rest, perceiving their fellows in this distress, fled into the high mountains. Two women, not being so apt to escape as the men were, the one for her age, and the other being encumbered with a young child, we took. The old wretch, whom divers of our sailors supposed to be either a devil or a witch, had her buskins plucked off to see if she were cloven-footed, and for her ugly hue and deformity we let her go; the young woman and the child we brought away. We named the place where they were slain Bloody Point, and the bay or harbour Yorke's Sound, after the name of one of the captains of the two barques.

Having this knowledge both of their fierceness and cruelty, and perceiving that fair means as yet is not able to allure them to familiarity, we disposed ourselves, contrary to our inclination, something to be cruel, returned to their tents, and made a spoil of the same, where we found an old shirt, a doublet, a girdle, and also shoes of our men, whom we lost the year before; on nothing else unto them belonging could we set our eyes.

Their riches are not gold, silver, or precious drapery, but their said tents and boats made of the skins of red deer and seal skins, also dogs like unto wolves, but for the most part black, with other trifles, more to be wondered at for their strangeness than for any other commodity needful for our use.

Thus returning to our ship the 3rd of August, we departed from the west shore, supposed firm with America, after we had anchored there thirteen days, and so the 4th thereof we came to our general on the east shore, and anchored in a fair harbour named Anne Warwick's Sound, and to which is annexed an island, both named after the Countess of Warwick—Anne Warwick's Sound and Isle.

In this isle our general thought good for this voyage to freight both the ships and barques with such stone or gold mineral as he judged to countervail the charges of his first and this his second navigation to these countries, with sufficient interest to the venturers whereby they might both be satisfied for this time and also in time to come (if it please God and our prince) to expect a much more benefit out of the bowels of those septentrional parallels, which long time hath concealed itself till at this present, through the wonderful diligence and great danger of our general and others, God is contented with the revealing thereof. It riseth so abundantly, that from the beginning of August to the 22nd thereof (every man following the diligence of our general) we raised above ground 200 ton, which we judged a reasonable freight for the ship and two barques in the said Anne Warwick's Isle.

In the meantime of our abode here some of the country people came to show themselves unto us sundry times from the main shore, near adjacent to the said isle. Our general, desirous to have some news of his men whom he lost the year before, with some company with him repaired with the ship boat to commune or sign with them for familiarity, whereunto he is persuaded to bring them. They at the first show made tokens that three of his five men were alive, and desired pen, ink, and paper, and that within three or four days they would return, and, as we judged, bring those of our men which were living with them.

They also made signs or tokens of their king, whom they called Cacough, and how he was carried on men's shoulders, and a man far surmounting any of our company in bigness and stature.

With these tokens and signs of writing, pen, ink, and paper were delivered them, which they would not take at our hands, but being laid upon the shore, and the party gone away, they took up; which likewise they do when they desire anything for change of theirs, laying for that which is left so much as they think will countervail the same, and not coming near together. It seemeth they have been used to this trade or traffic with some other people adjoining, or not far distant from their country.

After four days some of them showed themselves upon the firm land, but not where they were before. Our general, very glad thereof, supposing to hear of our men, went from the island with the boat and sufficient company with him. They seemed very glad, and allured him about a certain point of the land, behind which they might perceive a company of the crafty villains to lie lurking, whom our general would not deal withal, for that he knew not what company they were, so with few signs dismissed them and returned to his company.

Another time, as our said general was coasting the country with two little pinnaces, whereby at our return he might make the better relation thereof, three of the crafty villains with a white skin allured us to them. Once again our general, for that he hoped to hear of his men, went towards them; at our coming near the shore whereon they were we might perceive a number of them lie hidden behind great stones, and those three in sight labouring by all means possible that some would come on land; and perceiving we made no haste, by words nor friendly signs, which they used by clapping their hands, and being without weapon, and but three in sight, they sought further means to provoke us thereunto. One alone laid flesh on the shore, which we took up with the boat-hook as necessary victuals for the relieving of the man, woman, and child whom we had taken, for that as yet they could not digest our meat; whereby they perceived themselves deceived of their expectation for all their crafty allurements. Yet once again to make, as it were, a full show of their crafty natures and subtle sleights, to the intent thereby to have entrapped and taken some of our men, one of them counterfeited himself impotent and lame of his legs, who seemed to descend to the water's side with great difficulty, and to cover his craft the more one of his fellows came down with him, and in such places where he seemed unable to pass, he took him on his shoulders, set him by the water's side, and departed from him, leaving him, as it should seem, all alone; who, playing his counterfeit pageant very well, thought thereby to provoke some of us to come on shore, not fearing but that one of us might make our party good with a lame man.

Our general, having compassion of his impotency, thought good, if it were possible, to cure him thereof; wherefore he caused a soldier to shoot at him with his calever, which grazed before his face. The counterfeit villain deliverly fled without any impediment at all, and got him to his bow and arrows, and the rest from their lurking holes with their weapons, bows, arrows, slings, and darts. Our general caused some calevers to be shot off at them, whereby, some being hurt, they might hereafter stand in more fear of us.

This was all the answer for this time we could have of our men, or of our general's letter. Their crafty dealing at these three several times being thus manifest unto us, may plainly show their disposition in other things to be correspondent. We judged that they used these stratagems thereby to have caught some of us for the delivering of the man, woman, and child, whom we had taken.

They are men of a large corporature, and good proportion; their colour is not much unlike the sunburnt countryman, who laboureth daily in sun for his living.

They wear their hair something long, and cut before either with stone or knife, very disorderly. Their women wear their hair long, knit up with two loops, showing forth on either side of their faces, and the rest faltered upon a knot. Also, some of their women tint their faces proportionally, as chin, cheeks, and forehead and the wrists of their hands, whereupon they lay a colour which continueth dark azurine.

They eat their meat all raw, both flesh, fish, and fowl, or something parboiled with blood, and a little water, which they drink. For lack of water, they will eat ice that is hard frozen as pleasantly as we will do sugar-candy, or other sugar.

If they, for necessity's sake, stand in need of the premises, such grass as the country yieldeth they pluck up and eat, not daintily, or saladwise, to allure their stomachs to appetite, but for necessity's sake, without either salt, oils, or washing, like brute beasts devouring the same. They neither use table, stool, or table-cloth for comeliness: but when they are imbrued with blood, knuckle deep, and their knives in like sort, they use their tongues as apt instruments to lick them clean; in doing whereof they are assured to lose none of their victuals.

They keep certain dogs, not much unlike wolves, which they yoke together, as we do oxen and horses, to a sled or trail, and so carry their necessaries over the ice and snow, from place to place, as the captain, whom we have, made perfect signs. And when those dogs are not apt for the same use, or when with hunger they are constrained for lack of other victuals, they eat them, so that they are as needful for them, in respect of their bigness, as our oxen are for us.

They apparel themselves in the skins of such beasts as they kill, sewed together with the sinews of them. All the fowl which they kill they skin, and make thereof one kind of garment or other to defend them from the cold.

They make their apparel with hoods and tails, which tails they give, when they think to gratify any friendship shown unto them; a great sign of friendship with them. The men have them not so syde as the women.

The men and women wear their hose close to their legs, from the waist to the knee, without any open before, as well the one kind as the other. Upon their legs they wear hose of leather, with the fur side inward, two or three pair on at once, and especially the women. In those hose they put their knives, needles, and other things needful to bear about. They put a bone within their hose, which reacheth from the foot to the knee, whereupon they draw their said hose, and so in place of garters they are holden from falling down about their feet.

They dress their skins very soft and supple with the hair on. In cold weather or winter they wear the fur side inward, and in summer outward. Other apparel they have none but the said skins.

Those beasts, fishes, and fowls which they kill are their meat, drink, apparel, houses, bedding, hose, shoes, thread, and sails for their boats, with many other necessaries, whereof they stand in need, and almost all their riches.

The houses are tents made of seal skins, pitched up with four fir quarters, four-square, meeting at the top, and the skins sewed together with sinews, and laid thereupon; they are so pitched up, that the entrance into them is always south, or against the sun.

They have other sort of houses, which we found not to be inhabited, which are raised with stones and whalebones, and a skin laid over them to withstand the rain, or other weather; the entrance of them being not much unlike an oven's mouth, whereunto, I think, they resort for a time to fish, hunt, and fowl, and so leave them until the next time they come thither again.

Their weapons are bows, arrows, darts, and slings. Their bows are of wood, of a yard long, sinewed on the back with firm sinews, not glued to, but fast girded and tied on. Their bow strings are likewise sinews. Their arrows are three pieces, nocked with bone and ended with bone; with those two ends, and the wood in the midst, they pass not in length half a yard, or little more. They are feathered with two feathers, the pen end being cut away, and the feathers laid upon the arrow with the broad side to the wood, insomuch, that they seem, when they are tied on, to have four feathers. They have likewise three sorts of heads to those arrows; one sort of stone or iron, proportioned like to a heart; the second sort of bone much like unto a stopt head, with a hook on the same, the third sort of bone likewise, made sharp at both sides, and sharp pointed. They are not made very fast, but lightly tied to, or else set in a nocke, that, upon small occasion, the arrow leaveth these heads behind them; they are of small force except they be very near when they shoot.

Their darts are made of two sorts: the one with many forks of bones in the fore end, and likewise in the midst; their proportions are not much unlike our toasting-irons, but longer; these they cast out of an instrument of wood very readily. The other sort is greater than the first aforesaid, with a long bone made sharp on both sides, not much unlike a rapier, which I take to be their most hurtful weapon.

They have two sorts of boats made of leather, set out on the inner side with quarters of wood, artificially tied together with thongs of the same; the greater sort are not much unlike our wherries, wherein sixteen or twenty men may sit; they have for a sail dressed the guts of such beasts as they kill, very fine and thin, which they sew together; the other boat is but for one man to sit and row in, with one oar.

Their order of fishing, hunting, and fowling, are with these said weapons; but in what sort or how they use them we have no perfect knowledge as yet.

I can suppose their abode or habitation not to be here, for that neither their houses nor apparel are of such force to withstand the extremity of cold that the country seemeth to be infected withal; neither do I see any sign likely to perform the same.

Those houses, or rather dens, which stand there, have no sign of footway, or anything else trodden, which is one of the chiefest tokens of habitation. And those tents, which they bring with them, when they have sufficiently hunted and fished, they remove to other places; and when they have sufficiently stored them of such victuals as the country yieldeth, or bringeth forth, they return to their winter stations or habitations. This conjecture do I make for the infertility which I perceive to be in that country.

They have some iron, whereof they make arrow-heads, knives, and other little instruments, to work their boats, bows, arrows, and darts withal, which are very unapt to do anything withal, but with great labour.

It seemeth that they have conversation with some other people, of whom for exchange they should receive the same. They are greatly delighted with anything that is bright or giveth a sound.

What knowledge they have of God, or what idol they adore, we have no perfect intelligence. I think them rather *anthropophagi*, or devourers of man's flesh, than otherwise; that there is no flesh or fish which they find dead (smell it never so filthily), but they will eat it as they find it without any other dressing. A loathsome thing, either to the beholders or the hearers. There is no manner of creeping beast hurtful, except some spiders (which as many affirm are signs of great store of gold), and also certain stinging gnats, which bite so fiercely that the place where they bite shortly after swelleth, and itcheth very sore.

They make signs of certain people that wear bright plates of gold in their foreheads and other places of their bodies.

The countries on both sides the straits lie very high, with rough stony mountains, and great quantity of snow thereon. There is very little plain ground, and no grass except a little, which is much like unto moss that groweth on soft ground, such as we get turfs in. There is no wood at all. To be brief, there is nothing fit or profitable for the use of man which that country with root yieldeth or bringeth forth; howbeit there is great quantity of deer, whose skins are like unto asses, their heads or horns do far exceed, as well in length as also in breadth, any in these our parts or countries: their feet likewise are as great as our oxen's, which we measure to be seven or eight inches in breadth. There are also hares, wolves, fishing bears, and sea-fowl of sundry sorts.

As the country is barren and unfertile, so are they rude, and of no capacity to culture the same to any perfection; but are contented by their hunting, fishing, and fowling, with raw flesh and warm blood, to satisfy their greedy paunches, which is their only glory.

There is great likelihood of earthquakes or thunder, for there are huge and monstrous mountains, whose greatest substance are stones, and those stones so shapen with some extraordinary means, that one is separated from another, which is discordant from all other quarries.

There are no rivers or running springs, but such as through the heat of the sun, with such water as descendeth from the mountains and hills, whereon great drifts of snow do lie, are engendered.

It argueth also that there should be none; for that the earth, which with the extremity of the winter is so frozen within, that that water which should have recourse within the same to maintain springs hath not his motion, whereof great waters have their origin, as by experience is seen otherwise. Such valleys as are capable to receive the water, that in the summer time, by the operation of the sun, descendeth from great abundance of snow, which continually lieth on the mountains, and hath no passage, sinketh into the earth, and so vanisheth away, without any runnel above the earth, by which occasion or continual standing of the said water the earth is opened and the great frost yieldeth to the force thereof, which in other places, four or five fathoms within the ground, for lack of the said moisture, the earth even in the very summer time is frozen, and so combineth the stones together, that scarcely instruments with great force can unknit them.

Also, where the water in those valleys can have no such passage away, by the continuance of time in such order as is before rehearsed, the yearly descent from the mountains filleth them full, that at the lowest bank of the same they fall into the next valley, and so continue as fishing ponds, in summer time full of water, and in the winter hard frozen, as by scars that remain thereof in summer may easily be perceived; so that the heat of summer is nothing comparable or of force to dissolve the extremity of cold that cometh in winter.

Nevertheless, I am assured, that below the force of the frost, within the earth, the waters have recourse, and empty themselves out of sight into the sea, which, through the extremity of the frost, are constrained to do the same; by which occasion, the earth within is kept the warmer, and springs have their recourse, which is the only nutriment of gold and minerals within the same.

There is much to be said of the commodities of these countries, which are couched within the bowels of the earth, which I let pass till more perfect trial be made thereof.

Thus conjecturing, till time, with the earnest industry of our general and others (who, by all diligence, remain pressed to explore the truth of that which is unexplored, as he hath to his everlasting praise found out that which is like to yield an innumerable benefit to his prince and country), offer further trial, I conclude.

The 23rd August, after we had satisfied our minds with freight sufficient for our vessels, though not our covetous desires, with such knowledge of the country, people, and other commodities as are before rehearsed, the 24th thereof we departed there hence: the 17th of September we fell with the Land's End of England, and so to Milford Haven, from whence our general rowed to the court for order to what port or haven to conduct the ship.

We lost our two barques in the way homeward, the one the 29th of August, the other the 31st of the same month, by occasion of great tempest and fog; howbeit, God restored the one to Bristol, and the other making his course by Scotland to Yarmouth. In this voyage we lost two men, one in the way by God's visitation, and the other homeward, cast overboard with a surge of the sea.

I could declare unto the readers the latitude and longitude of such places and regions as we have been at, but not altogether so perfectly as our masters and others, with many circumstances of tempests and other accidents incident to seafaring men, which seem not altogether strange, but I let them pass to their reports as men most apt to set forth and declare the same. I have also left the names of the countries on both the shores untouched for lack of understanding the people's language, as also for sundry respects not needful as yet to be declared.

Countries new explored, where commodity is to be looked for, do better accord with a new name given by the explorers than an uncertain name by a doubtful author.

Our general named sundry islands, mountains, capes, and harbours after the names of divers noblemen, and other gentlemen his friends, as well on the one shore as also on the other.

THE THIRD AND LAST VOYAGE INTO META INCOGNITA,

Made by Master Martin Frobisher, in the year 1578, *written by Thomas Ellis.*

These are to let you know, that upon the 25th May, the *Thomas Allen*, being vice-admiral, whose captain was Master Yorke; Master Gibbes, master; Master Christopher Hall, pilot, accompanied with the rear-admiral, named the *Hopewell*, whose captain was Master Henry Carew, the Master Andrew Dier, and certain other ships, came to Gravesend, where we anchored, and abode the coming of certain other of our fleet, which were not yet come.

The 27th of the same month, our fleet being now come together, and all things pressed in a readiness, the wind favouring and tide serving, we being of sails in number eight, weighed anchors, and hoisted our sails towards Harwich, to meet with our admiral and the residue, which then and there abode our arrival, where we safely arrived the 28th thereof; finding there our admiral, whom we, with the discharge of certain pieces, saluted (according to order and duty), and were welcomed with the like courtesy, which being finished we landed, where our general continued mustering his soldiers and miners, and setting things in order appertaining to the voyage, until the last of the said month of May, which day we hoisted our sails, and committing ourselves to the conducting of Almighty God, we set forward toward the West Country, in such lucky wise and good success, that by the 5th June we passed the Dursies, being the utmost part of Ireland, to the westward.

And here it were not much amiss, nor far from our purpose, if I should a little discourse and speak of our adventures and chances by the way, as our landing at Plymouth, as also the meeting of certain poor men, which were robbed and spoiled of all that they had by pirates and rovers; amongst whom was a man of Bristol, on whom our general used his liberality, and sent him away with letters into England.

But because such things are impertinent to the matter, I will return (without any more mentioning of the same) to that from which I have digressed and swerved, I mean our ships, now sailing on the surging seas, sometimes passing at pleasure with a wished eastern wind, sometimes hindered of our course again by the western blasts, until the 20th day of the foresaid month of June, on which day in the morning we fell in with Friesland, which is a very high and cragged land, and was almost clean covered with snow, so that we might see nought but craggy rocks and the tops of high and huge hills, sometimes (and for the most part) all covered with foggy mists. There might we also perceive the great isles of ice lying on the seas like mountains, some small, some big, of sundry kinds of shapes, and such a number of them, that we could not come near the shore for them.

Thus sailing along the coast, at the last we saw a place somewhat void of ice, where our general (accompanied with certain other) went ashore, where they saw certain tents made of beasts' skins, and boats much like unto theirs of Meta Incognita. The tents were furnished with flesh, fish, skins, and other trifles: amongst the which was found a box of nails, whereby we did conjecture that they had either artificers amongst them, or else a traffic with some other nation. The men ran away, so that we could have no conference or communication with them. Our general (because he would have them no more to flee, but rather encouraged to stay through his courteous dealing) gave commandment that his men should take nothing away with them, saving only a couple of white dogs, for which he left pins, points, knives, and other trifling things, and departed, without taking or hurting anything, and so came aboard, and hoisted sails and passed forwards.

But being scarce out of the sight thereof, there fell such a fog and hideous mist that we could not see one another; whereupon we struck our drums, and sounded our trumpets to the end we might keep together; and so continued all that day and night, till the next day, that the mist brake up; so that we might easily perceive all the ships thus sailing together all that day, until the next day, being the 22nd of the same, on which day we saw an infinite number of ice, from the which we cast about to shun the danger thereof.

But one of our small barques named the *Michael*, whose captain was Master Kinderslie, the master, Bartholomew Bull, lost our company, insomuch that we could not obtain the sight of her many days after, of whom I mean to speak further anon, when occasion shall be ministered, and opportunity served. Thus we continued on our course until the 2nd of July, on which day we fell with the Queen's Foreland, where we saw so much ice, that we thought it impossible to get into the straits, yet at the last we gave the adventure, and entered the ice.

Being in amongst it, we saw the *Michael*, of whom I spake before, accompanied with the, *Judith*, whose captain was Master Fenton, the master, Charles Jackman, bearing into the aforesaid ice, far distant from us, who in a storm that fell that present night (whereof I will at large, God willing, discourse hereafter), were severed from us, and being in, wandered up and down the straits amongst the ice, many days in great peril, till at the last (by the providence of God) they came safely to harbour in their wished port in the Countess of Warwick's Sound the 20th July aforesaid, ten days before any of the other ships; who going on shore, found where the people of the country had been, and had hid their provision in great heaps of stone, being both of flesh and fish, which they had killed, whereof we also found great store in other places after our arrival. They found also divers engines, as bows, slings, and darts. They found likewise certain pieces of the pinnace which our general left there the year before; which pinnace he had sunk, minding to have it again the next year.

Now, seeing I have entreated so much of the *Judith* and the *Michael*, I will return to the rest of the other ships, and will speak a little of the storm which fell, with the mishaps that we had, the night that we put into the ice, whereof I made mention before.

At the first entry into the ice, in the mouth of the straits, our passage was very narrow and difficult; but being once gotten in, we had a fair, open place without any ice for the most part; being a league in compass, the ice being round about us, and enclosing us, as it were, within the pales of a park. In which place (because it was almost night) we minded to take in our sails and lie a hull all that night. But the storm so increased, and the waves began to mount aloft, which brought the ice so near us, and coming in so fast upon us, that we were fain to bear in and out, where ye might espy an open place. Thus the ice coming on us so fast we were in great danger, looking every hour for death, and thus passed we on in that great danger, seeing both ourselves and the rest of our ships so troubled and tossed amongst the ice, that it would make the strongest-heart to relent.

At the last, the barque *Dionyse*, being but a weak ship, and bruised afore amongst the ice, being so leak that she no longer could carry above water, sank without saving any of the goods which were in her: the sight so abashed the whole fleet, that we thought verily we should have tasted of the same sauce. But nevertheless, we seeing them in such danger, manned our boats, and saved all the men, in such wise that not one perished. (God be thanked.)

The storm still increased and the ice enclosed us, that we were fain to take down top and topmasts; for the ice had so environed us, that we could see neither land nor sea as far as we could ken; so that we were fain to cut our cables to hang overboard for fenders, somewhat to ease the ship's sides from the great and dreary strokes of the ice; some with capstan bars, some fending off with oars, some with planks of two inches thick, which were broken immediately with the force of the ice, some going out upon the ice, to bear it off with their shoulders from the ships. But the rigorousness of the tempest was such, and the force of the ice so great, that not only they burst and spoiled the foresaid provision, but likewise so raised the sides of the ships that it was pitiful to behold, and caused the hearts of many to faint.

Thus continued we all that dismal and lamentable night, plunged in this perplexity, looking for instant death; but our God (who never leaveth them destitute which faithfully call upon Him), although He often punisheth for amendment's sake, in the morning caused the winds to cease, and the fog, which all that night lay on the face of the water, to clear, so that we might perceive about a mile from us a certain place clear from any ice, to the which with an easy breath of wind, which our God sent us, we bent ourselves, and furthermore He provided better for us than we deserved, or hoped for; for when we were in the foresaid clear place, He sent us a fresh gale at west, or at west-south-west, which set us clear without all the ice. And further He added more, for He sent us so pleasant a day, as the like we had not of a long time before, as after punishment consolation.

Thus we joyful whites, being at liberty, took in all our sails, and lay a hull, praising God for our deliverance, and stayed to gather together our fleet; which once being done, we seeing that none of them had any great hurt, neither any of them wanted, saving only they of whom I spake before, and the ship which was lost, then at the last we hoisted our sails, and lay bulting off and on, till such time as it would please God to take away the ice, that we might get into the straits.

As we thus lay off and on, we came by a marvellous huge mountain of ice, which surpassed all the rest that ever we saw, for we judged it to be near four score fathoms above water, and we thought it to be aground for anything that we could perceive, being there nine score fathoms deep, and of compass about half a mile.

Also the fifth of July there fell a hideous fog and mist, that continued till the nineteenth of the same, so that one ship could not see another. Therefore we were fain to bear a small sail, and to observe the time, but there ran such a current of tide, that it set us to the north-west of the Queen's Forehand, the back side of all the straits, where (through the contagious fog having no sight either of sun or star) we scarce knew where we were. In this fog the 10th July we lost the company of the *Vice-Admiral*, the *Anne Francis*, the *Busse of Bridgewater*, and the *Francis of Foy*.

The sixteenth day, one of our small barques, named the *Gabriel*, was sent by our general to bear in with the land, to descry it, where, being on land, they met with the people of the country, which seemed very humane and civilised, and offered to traffic with our men, proffering them fowls and skins for knives and other trifles, whose courtesy caused us to think that they had small conversation with the other of the straits. Then we bare back again, to go with the Queen's Forehand, and the 18th day we came by two islands, whereon we went on shore, and found where the people had been, but we saw none of them. This day we were again in the ice, and like to be in as great peril as we were at the first. For through the darkness and obscurity of the foggy mist we were almost run on rocks and islands before we saw them: but God (even miraculously) provided for us, opening the fogs that we might see clearly, both where and in what danger we presently were, and also the way to escape; or else, without fail we had ruinously run upon the rocks.

When we knew perfectly our instant case, we cast about to get again on sea board, which (God be thanked) by might we obtained, and praised God. The clear continued scarce an hour, but the fog fell again as thick as ever it was.

Then the *Rear-Admiral* and the *Bear* got themselves clear without danger of ice and rocks, struck their sails and lay a hull, staying to have the rest of the fleet come forth, which as yet had not found the right way to clear themselves from the danger of rocks and ice, until the next morning, at what time the *Rear-Admiral* discharged certain warning pieces, to give notice that she had escaped, and that the rest (by following of her) might set themselves free, which they did that day. Then having gathered ourselves together, we proceeded on our purposed voyage, bearing off, and keeping ourselves distant from the coast, until the 19th day of July, at which time the fogs brake up and dispersed, so that we might plainly and clearly behold the pleasant air which had so long been taken from us by the obscurity of the foggy mists; and, after that time, we were not much encumbered therewith until we had left the confines of the country.

Then we, espying a fair sound, supposed it to go into the straits, between the Queen's Foreland and Jackman's Sound, which proved as we imagined. For our general sent forth again the *Gabriel* to discover it, who passed through with much difficulty, for there ran such an extreme current of a tide, with so horrible a gulf, that with a fresh gale of wind they were scarce able to stem it, yet at the length with great travel they passed it, and came to the straits, where they met with the *Thomas Allen*, the *Thomas of Ipswich*, and the *Busse of Bridgewater*, who all together adventured to bear into the ice again, to see if they could obtain their wished port. But they were so encumbered, that with much difficulty they were able to get out again, yet at the last they escaping the *Thomas Allen* and the *Gabriel*, bear in with the western shore, where they found harbour, and they moored their ships until the 4th of August, at which time they came to us, in the Countess of Warwick's Sound. The *Thomas of Ipswich* caught a great leak, which caused her to cast again to sea board, and so was mended.

We sailed along still by the coast until we came to the Queen's Forehand, at the point whereof we met with part of the gulf aforesaid, which place or gulf (as some of our masters do credibly report) doth flow nine hours and ebbs but three. At that point we discovered certain lands southward, which neither time nor opportunity would serve to search. Then being come to the mouth of the straits, we met with the *Anne Francis*, who had lain bulting up and down ever since her departure alone, never finding any of her company. We met then also the *Francis of Foy*, with whom again we intended to venture and get in, but the ice was yet so thick, that we were compelled again to retire and get us on sea board.

There fell also the same day, being the 26th July, such a horrible snow, that it lay a foot thick upon the hatches, which froze as fast as it fell.

We had also at other times divers cruel storms, both snow and hail, which manifestly declared the distemperature of the country: yet for all that we were so many times repulsed and put back from our purpose, knowing that lingering delay was not profitable for us, but hurtful to our voyage, we mutually consented to our valiant general once again to give the onset.

The 28th day, therefore, of the same July we assayed, and with little trouble (God be praised) we passed the dangers by daylight. Then night falling on the face of the earth, we hulled in the clear, till the cheerful light of the day had chased away the noisome darkness of the night, at which the we set forward toward our wished port; by the 30th day we obtained our expected desire, where we found the *Judith* and the *Michael*, which brought no small joy unto the general, and great consolation to the heavy hearts of those wearied wights.

The 30th day of July we brought our ships into the Countess of Warwick's Sound, and moored them, namely these ships, the *Admiral*, the *Rear-Admiral*, the *Francis of Foy*, the *Bear*, *Armenel*, the *Salomon*, and the *Busse of Bridgewater*, which being done, our general commanded us all to come ashore upon the Countess Island, where he set his miners to work upon the mine, giving charge with expedition to despatch with their lading.

Our general himself, accompanied with his gentleman, divers times made roads into sundry parts of the country, as well to find new mines as also to find out and see the people of the country. He found out one mine, upon an island by Bear's Sound, and named it the Countess of Sussex Island. One other was found in Winter's Fornace, with divers others, to which the ships were sent sunderly to be laden. In the same roads he met with divers of the people of the country at sundry times, as once at a place called David's Sound, who shot at our men, and very desperately gave them the onset, being not above three or four in number, there being of our countrymen above a dozen; but seeing themselves not able to prevail, they took themselves to flight, whom our men pursued, but being not used to such craggy cliffs, they soon lost the sight of them, and so in vain returned.

We also saw them at Bear's Sound, both by sea and land, in great companies; but they would at all times keep the water between them and us. And if any of our ships chanced to be in the sound (as they came divers times), because the harbour was not very good, the ship laded, and departed again; then so long as any ships were in sight, the people would not be seen. But when as they perceived the ships to be gone, they would not only show themselves standing upon high cliffs, and call us to come over unto them, but also would come in their boats very near to us, as it were to brag at us; whereof our general, having advertisement, sent for the captain and gentlemen of the ships to accompany and attend upon him, with the captain also of the *Anne Francis*, who was but the night before come unto us. For they and the fleet-boat, having lost us the 26th day, in the great snow, put into a harbour in the Queen's Forehand, where they found good ore, wherewith they laded themselves, and came to seek the general; so that now we had all our ships, saving one barque, which was lost, and the *Thomas of Ipswich* who (compelled by what fury I know not) forsook our company, and returned home without lading.

Our general, accompanied with his gentlemen (of whom I spake), came altogether to the Countess of Sussex Island, near to Bear's Sound, where he manned out certain pinnaces and went over to the people, who, perceiving his arrival, fled away with all speed, and in haste left certain darts and other engines behind them which we found, but the people we could not find.

The next morning our general, perceiving certain of them in boat upon the sea, gave chase to them in a pinnace under sail, with a fresh gale of wind, but could by no means come near unto them, for the longer he sailed the farther off he was from them, which well showed their cunning and activity. Thus time wearing away, and the day of our departure approaching, our general commanded to lade with all expedition, that we might be again on sea board with our ship; for whilst we were in the country we were in continual danger of freezing in, for often snow and hail, often the water was so much frozen and congealed in the night, that in the morning we could scarce row our boats or pinnaces, especially in Dier's Sound, which is a calm and still water, which caused our general to make the more haste, so that by the 30th day of August we were all laden, and made all things ready to depart. But before I proceed any further herein, to show what fortune befell at our departure, I will turn my pen a little to Master Captain Fenton, and those gentlemen which should have inhabited all the year in those countries, whose valiant minds were much to be commended, that neither fear of force, nor the cruel nipping storms of the raging winter, neither the intemperature of so unhealthful a country, neither the savageness of the people, neither the sight and show of such and so many strange meteors, neither the desire to return to their native soil, neither regard of friends, neither care of possessions and inheritances, finally, not the love of life (a thing of all other most sweet), neither the terror of dreadful death itself, might seem to be of sufficient force to withdraw their prowess, or to restrain from that purpose, thereby to have profited their country; but that with most willing hearts, venturous minds, stout stomachs, and singular manhood, they were content there to have tarried for the time, among a barbarous and uncivilised people, infidels and miscreants, to have made their dwelling, not terrified with the manifold and imminent dangers which they were like to run into; and seeing before their eyes so many casualties, whereto their life was subject, the least whereof would have made a milksop Thersites astonished and utterly discomfited; being, I say, thus minded and purposed, they deserved special commendation, for, doubtless, they had done as they intended, if luck had not withstood their willingness, and if that fortune had not so frowned upon their intents.

For the bark *Dionyse*, which was lost, had in her much of their house, which was prepared and should have been builded for them, with many other implements. Also the *Thomas of Ipswich*, which had most of their provision in her, came not into the straits at all, neither did we see her since the day we were separated in the great snow (of which I spake before). For these causes, having not their house nor yet provision, they were disappointed of their pretence to tarry, and therefore laded their ships and so came away with us.

But before we took shipping, we builded a little house in the Countess of Warwick's Island, and garnished it with many kinds of trifles, as pins, points, laces, glasses, combs, babes on horseback and on foot, with innumerable other such fancies and toys, thereby to allure and entice the people to some familiarity against other years.

Thus having finished all things we departed the country (as I said before); but because the *Busse* had not lading enough in her, she put into Bear's Sound to take a little more. In the meanwhile, the *Admiral*, and the rest without the sea, stayed for her. And that night fell such an outrageous tempest, beating on our ships with such vehement rigour that anchor and cable availed nought, for we were driven on rocks and islands of ice, insomuch that had not the great goodness of God been miraculously showed to us, we had been cast away every man. This danger was more doubtful and terrible than any that preceded or went before, for there was not any one ship (I think) that escaped without damage. Some lost anchor, and also gables, some boats, some pinnaces, some anchor, gables, boats, and pinnaces.

This boisterous storm so severed us one from another, that one ship knew not what was become of another. The *Admiral* knew not where to find the *Vice-Admiral* or *Rear-Admiral*, or any other ship of our company. Our general, being on land in Bear's Sound, could not come to his ship, but was compelled to go aboard the *Gabriel*, where he continued all the way homewards, for the boisterous blasts continued so extremely, and so long a time, that it sent us homeward (which was God's favour towards us), will we, nill we, in such haste, as not any one of us were able to keep in company of other, but were separated. And if by chance any one ship did overtake other by swiftness of sail, or met (as they often did), yet was the rigour of the wind so hideous, that they could not continue company together the space of one whole night.

Thus our journey outward was not so pleasant, but our coming thither, entering the coasts and country by narrow straits, perilous ice, and swift tides, our times of abode there in snow and storms, and our departure from thence, the 3rd of August, with dangerous blustering winds and tempest's, which that night arose, was as uncomfortable, separating us so, as we sailed, that not any of us met together until the 28th of September, which day we fell on the English coasts, between Scilly and the Land's End, and passed the Channel, until our arrival in the river Thames.

THE REPORT OF THOMAS WIARS,

Passenger in the "Emmanuel," otherwise called the "Busse of Bridgewater," wherein James Leeche was Master, one of the ships in the last voyage of Master Martin Frobisher, 1578, concerning the discovery of the great island in their way homeward, the 12th of September.

The *Busse of Bridgewater* was left in Bear's Sound, at Meta Incognita, the 2nd day of September, behind the fleet, in some distress, through much wind riding near the lee shore, and forced there to ride it out upon the hazard of her cables and anchors, which were all aground but two. The 3rd of September being fair weather, and the wind north-north-west, she set sail, and departed thence and fell with Friesland, on he 8th day of September, at six of the clock at night, and then they set off from the south-west point of Friesland, the wind being at east and east-south-east; but that night the wind veered southerly, and shifted oftentimes that night. But on the 10th day, in the morning, the wind at west-north-west, fair weather, they steered south-east and by south, and continued that course until the 12th day of September, when about 11 o'clock before noon they descried a land, which was from them about five leagues, and the southernmost part of it was south-east-by-east from them, and the northernmost next north-north-east, or north-east. The master accounted that Friesland, the south-east point of it, was from him at that instant, when he first descried this new island, north-west-by-north fifty leagues. They account this island to be twenty-five leagues long, and the longest way of it south-east and north-west. The southern part of it is in the latitude of fifty-seven degrees and one second part, or thereabout. They continued in sight of it from the twelfth day at eleven of the clock till the thirteenth day three of the clock in the afternoon, when they left it; and the last part they saw of it bare from them north-west-by-north. There appeared two harbours upon that coast, the greatest of them seven leagues to the northwards of the southernmost point, the other but four leagues. There was very much ice near the same land, and also twenty or thirty leagues from it, for they were not clear of ice till the 15th day of September, afternoon. They plied their voyage homeward, and fell with the west part of Ireland, about Galway, and had first sight of it on the 25th day of September.

THE FIRST VOYAGE OF MASTER JOHN DAVIS,

Undertaken in June, 1585, for the discovery of the North-West Passage, written by John James Marchant, servant to the Worshipful Master William Sanderson.

Certain honourable personages and worthy gentlemen of the Court and country, with divers worshipful merchants of London and of the West Countrie, moved with desire to advance God's glory, and to seek the good of their native country, consulting together of the likelihood of the discovery of the North-West Passage, which heretofore had been attempted, but unhappily given over by accidents unlooked for, which turned the enterprisers from their principal purpose, resolved, after good deliberation, to put down their adventures, to provide for necessary shipping, and a fit man to be chief conductor of this so hard an enterprise. The setting forth of this action was committed by the adventurers especially to the care of Master William Sanderson, merchant of London, who was so forward therein, that besides his travel, which was not small, he became the greatest adventurer with his purse, and commended unto the rest of the company one Master John Davis, a man very well grounded in the principles of the art of navigation, for captain and chief pilot of this exploit.

Thus, therefore, all things being put in a readiness, we departed from Dartmouth the 7th of June towards the discovery of the aforesaid North-West Passage with two barques, the one being of fifty tons, named the *Sunshine*, of London, and the other being thirty-five tons, named the *Moonshine*, of Dartmouth. In the *Sunshine* we had twenty-three persons, whose names are these following: Master John Davis, captain; William Eston, master; Richard Pope, master's mate; John Jane, merchant; Henry Davie, gunner; William Crosse, boatswain; John Bagge, Walter Arthur, Luke Adams, Robert Coxworthie, John Ellis, John Kelly, Edward Helman, William Dicke, Andrew Maddocke, Thomas Hill, Robert Wats, carpenter, William Russell, Christopher Gorney, boy; James Cole, Francis Ridley, John Russel, Robert Cornish, musicians.

The *Moonshine* had nineteen persons, William Bruton, captain; John Ellis, master; the rest mariners.

The 7th of June the captain and the master drew out a proportion for the continuance of our victuals.

The 8th day, the wind being at south-west and west-south-west, we put in for Falmouth, where we remained until the 13th.

The 13th the wind blew at north, and being fair weather we departed.

The 14th, with contrary wind, we were forced to put into Scilly.

The 15th we departed thence, having the wind north and by east, moderate and fair weather.

The 16th we were driven back again, and were constrained to arrive at New Grimsby, at Scilly; here the wind remained contrary twelve days, and in that space the captain, the master, and I went about all the islands, and the captain did plan out and describe the situation of all the islands, rocks, and harbours to the exact use of navigation, with lines and scale thereunto convenient.

The 28th, in God's name, we departed, the wind being easterly, but calm.

The 29th very foggy.

The 30th foggy.

The 1st of July we saw great store of porpoises, the master called for a harping-iron, and shot twice or thrice; sometimes he missed, and at last shot one and struck him in the side, and wound him into the ship; when we had him aboard, the master said it was a darley head.

The 2nd we had some of the fish boiled, and it did eat as sweet as any mutton.

The 3rd we had more in sight, and the master went to shoot at them, but they were so great, that they burst our irons, and we lost both fish, irons, pastime, and all; yet, nevertheless, the master shot at them with a pike, and had well-nigh gotten one, but he was so strong, that he burst off the bars of the pike and went away. Then he took the boat-hook, and hit one with that; but all would not prevail, so at length we let them alone.

The 6th we saw a very great whale, and every day after we saw whales continually.

The 16th, 17th, and 18th we saw great store of whales.

The 19th of July we fell into a great whirling and brustling of a tide, setting to the northward; and sailing about half a league we came into a very calm sea, which bent to the south-south-west. Here we heard a mighty great roaring of the sea, as if it had been the breach of some shore, the air being so foggy and full of thick mist, that we could not see the one ship from the other, being a very small distance asunder; so the captain and the master, being in distrust how the tide might set them, caused the *Moonshine* to hoist out her boat and to sound, but they could not find ground in three hundred fathoms and better. Then the captain, master, and I went towards the breach to see what it should be, giving charge to our gunners that at every blast they should shoot off a musket shot, to the intent we might keep ourselves from losing them; then coming near to the breach, we met many islands of ice floating, which had quickly compassed us about. Then we went upon some of them, and did perceive that all the roaring which we heard was caused only by the rolling of this ice together. Our company seeing us not to return according to our appointment, left off shooting muskets and began to shoot falconets, for they feared some mishap had befallen us; but before night we came aboard again, with our boat laden with ice, which made very good fresh water. Then we bent our course toward the north, hoping by that means to double the land.

The 20th, as we sailed along the coast, the fog brake up, and we discovered the land, which was the most deformed, rocky, and mountainous land that ever we saw, the first sight whereof did show as if it had been in form of a sugar loaf, standing to our sight above the clouds, for that it did show over the fog like a white liste in the sky, the tops altogether covered with snow, and the shore beset with ice a league off into the sea, making such irksome noise as that it seemed to be the true pattern of desolation, and after the same our captain named it the land of desolation.

The 21st the wind came northerly and overblew, so that we were constrained to bend our course south again, for we perceived that we were run into a very deep bay, where we were almost compassed with ice, for we saw very much towards the north-north-east, west, and south-west; and this day and this night we cleared ourselves of the ice, running south-south-west along the shore.

Upon Thursday, being the 22nd of this month, about three of the clock in the morning, we hoisted out our boat, and the captain, with six sailors, went towards the shore, thinking to find a landing-place, for the night before we did perceive the coast to be void of ice to our judgment; and the same night we were all persuaded that we had seen a canoe rowing along the shore, but afterwards we fell in some doubt of it, but we had no great reason so to do. The captain, rowing towards the shore, willed the master to bear in with the land after him; and before he came near the shore, by the space of a league, or about two miles, he found so much ice that he could not get to land by any means. Here our mariners put to their lines to see if they could get any fish, because there were so many seals upon the coast, and the birds did beat upon the water, but all was in vain: the water about this coast was very black and thick, like to a filthy standing pool; we sounded, and had ground in 120 fathoms. While the captain was rowing to the shore our men saw woods upon the rocks, like to the rocks of Newfoundland, but I could not discern them; yet it might be so very well, for we had wood floating upon the coast every day, and the *Moonshine* took up a tree at sea not far from the coast, being sixty foot of length and fourteen handfuls about, having the root upon it. After, the captain came aboard, the weather being very calm and fair, we bent our course toward the south with intent to double the land.

The 23rd we coasted the land which did lie east-north-east and west-south-west.

The 24th, the wind being very fair at east, we coasted the land, which did lie east and west, not being able to come near the shore by reason of the great quantity of ice. At this place, because the weather was somewhat cold by reason of the ice, and the better to encourage our men, their allowance was increased. The captain and the master took order that every mess, being five persons, should have half a pound of bread and a can of beer every morning to breakfast. The weather was not very cold, but the air was moderate, like to our April weather in England. When the wind came from the land or the ice it was somewhat cold, but when it came off the sea it was very hot.

The 25th of this month we departed from sight of this land at six of the clock in the morning, directing our course to the north-westward, hoping in God's mercy to find our desired passage, and so continued above four days.

The 29th of July we discovered land in 64 degrees 15 minutes of latitude, bearing north-east from us. The wind being contrary to go to the north-westward, we bear in with this land to take some view of it, being utterly void of the pester of ice, and very temperate. Coming near the coast we found many fair sounds and good roads for shipping, and many great inlets into the land, whereby we judged this land to be a great number of islands standing together. Here, having moored our barque in good order, we went on shore upon a small island to seek for water and wood. Upon this island we did perceive that there had been people, for we found a small shoe and pieces of leather sewed with sinews and a piece of fur, and wool like to beaver. Then we went upon another island on the other side of our ships, and the captain, the master, and I, being got up to the top of a high rock, the people of the country having espied us made a lamentable noise, as we thought, with great outcries and screechings; we, hearing them, thought it had been the howling of wolves. At last I halloed again, and they likewise cried; then we, perceiving where they stood—some on the shore, and one rowing in a canoe about a small island fast by them—we made a great noise, partly to allure them to us and partly to warn our company of them. Whereupon Master Bruton and the master of his ship, with others of their company, made great haste towards us, and brought our musicians with them from our ship, purposing either by force to rescue us, if needs should so require, or with courtesy to allure the people. When they came unto us we caused our musicians to play, ourselves dancing and making many signs of friendship. At length there came ten canoes from the other islands, and two of them came so near the shore where we were that they talked with us, the other being in their boats a pretty way off. Their pronunciation was very hollow through the throat, and their speech such as we could not understand, only we allured them by friendly embracings and signs of courtesy. At length one of them, pointing up to the sun with his hand, would presently strike his breast so hard that we might hear the blow. This he did many times before he would any way trust us. Then John Ellis, the master of the *Moonshine*, was appointed to use his best policy to gain their friendship, who shook his breast and pointed to the sun after their order, which when he had divers times done they began to trust him, and one of them came on shore, to whom we threw our caps, stockings, and gloves, and such other things as then we had about us, playing with our music, and making signs of joy, and dancing. So the night coming we bade them farewell, and went aboard our barques.

The next morning, being the 30th of July, there came thirty-seven canoes rowing by our ships calling to us to come on shore; we not making any great haste unto them, one of them went up to the top of the rock, and leaped and danced as they had done the day before, showing us a seal skin, and another thing made like a timbrel, which he did beat upon with a stick, making a noise like a small drum. Whereupon we manned our boats and came to them, they all staying in their canoes. We came to the water's side, where they were, and after we had sworn by the sun after their fashion they did trust us. So I shook hands with one of them, and he kissed my hand, and we were very familiar with them. We were in so great credit with them upon this single acquaintance that we could have anything they had. We bought five canoes of them; we bought their clothes from their backs, which were all made of seal skins and birds' skins; their buskins, their hose, their gloves, all being commonly sewed and well dressed, so that we were fully persuaded that they have divers artificers among them. We had a pair of buskins of them full of fine wool like beaver. Their apparel for heat was made of birds' skins with their feathers on them. We saw among them leather dressed like glover's leather, and thick thongs like white leather of good length. We had of their darts and oars, and found in them that they would by no means displease us, but would give us whatsoever we asked of them, and would be satisfied with whatsoever we gave them. They took great care one of another, for when we had bought their boats then two other would come, and carry him away between them that had sold us his. They are a very tractable people, void of craft or double dealing, and easy to be brought to any civility or good order, but we judged them to be idolaters, and to worship the sun.

During the time of our abode among these islands we found reasonable quantity of wood, both fir, spruce, and juniper; which, whether it came floating any great distance to these places where we found it, or whether it grew in some great islands near the same place by us not yet discovered, we know not. But we judge that it groweth there farther into the land than we were, because the people had great store of darts and oars which they made none account of, but gave them to us for small trifles as points and pieces of paper. We saw about this coast marvellous great abundance of seals sculling together like sculls of small fish. We found no fresh water among these islands, but only snow-water, whereof we found great pools. The cliffs were all of such ore as Master Frobisher brought from Meta Incognita. We had divers shewes of study or Moscovie glass, shining not altogether unlike to crystal. We found an herb growing upon the rocks whose fruit was sweet, full of red juice, and the ripe ones were like currants. We found also birch and willow growing like shrubs low to the ground. These people have great store of furs as we judged. They made shows unto us the 30th of this present, which was the second time of our being with them, after they perceived we would have skins and furs, that they would go into the country and come again the next day with such things as they had; but this night the wind coming fair the captain and the master would by no means detract the purpose our discovery. And so the last of this month, about four of the clock in the morning, in God's name we set sail, and were all that day becalmed upon the coast.

The 1st of August we had a fair wind, and so proceeded towards the northwest for our discovery.

The 6th of August we discovered land in 66 degrees 40 minutes of latitude altogether void from the pester of ice; we anchored in a very fair road, under a very brave mount, the cliffs whereof were as orient as gold. This mount was named Mount Raleigh; the road where our ships lay at anchor was called Totnes Road; the sound which did compass the mount was named Exeter Sound; the foreland towards the north was called Dier's Cape; the foreland towards the south was named Cape Walsingham. So soon as we were come to an anchor in Totnes Road under Mount Raleigh we espied four white bears at the foot of the mount. We, supposing them to be goats or wolves, manned our boats and went towards them, but when we came near the shore we found them to be white bears of a monstrous bigness; we, being desirous of fresh victual and the sport, began to assault them, and I being on land, one of them came down the hill right against me. My piece was charged with hail-shot and a bullet; I discharged my piece and shot him in the neck; he roared a little, and took the water straight, making small account of his hurt. Then we followed him with our boat, and killed him with boars' spears, and two more that night. We found nothing in their maws, but we judged by their dung that they fed upon grass, because it appeared in all respects like the dung of a horse, wherein we might very plainly see the very straws.

The 7th we went on shore to another bear, which lay all night upon the top of an island under Mount Raleigh, and when we came up to him he lay fast asleep. I levelled at his head, and the stone of my piece gave no fire; with that he looked up and laid down his head again; then I shot, being charged with two bullets, and struck him in the head; he, being but amazed, fell backwards, whereupon we ran all upon him with boar spears and thrust him in the body, yet for all that he gripped away our boar spears and went towards the water, and as he was going down he came back again. Then our master shot his boar spear and struck him in the head, and made him to take the water, and swim into a cove fast by, where we killed him and brought him aboard. The breadth of his fore foot from one side to the other was fourteen inches over. They were very fat, so as we were constrained to cast the fat away. We saw a raven upon Mount Raleigh. We found withies, also, growing low like shrubs, and flowers like primroses in the said place. The coast is very mountainous, altogether without wood, grass, or earth, and is only huge mountains of stone, but the bravest stone that ever we saw. The air was very moderate in this country.

The 8th we departed from Mount Raleigh, coasting along the shore which lieth south-south-west and east-north-east.

The 9th our men fell in dislike of their allowance because it was so small as they thought. Whereupon we made a new proportion, every mess, being five to a mess, should have four pound of bread a day, twelve wine quarts of beer, six new land fishes, and the flesh days a gin of pease more; so we restrained them from their butter and cheese.

The 11th we came to the most southerly cape of this land, which we named the Cape of God's Mercy, as being the place of our first entrance for the discovery. The weather being very foggy we coasted this north land; at length when it brake up we perceived that we were shot into a very fair entrance or passage, being in some places twenty leagues broad and in some thirty, altogether void of any pester of ice, the weather very tolerable, and the water of the very colour, nature, and quality of the main ocean, which gave us the greater hope of our passage. Having sailed north-west sixty leagues in this entrance, we discovered certain islands standing in the midst thereof, having open passages on both sides. Whereupon our ships divided themselves, the one sailing on the north side, the other on the south side of the said isles, where we stayed five days, having the wind at south-east, very foggy, and foul weather.

The 14th we went on shore and found signs of people, for we found stones laid up together like a wall, and saw the skull of a man or a woman.

The 15th we heard dogs howl on the shore, which we thought had been wolves, and therefore we went on shore to kill them. When we came on land the dogs came presently to our boat very gently, yet we thought they came to prey upon us, and therefore we shot at them and killed two, and about the neck of one of them we found a leathern collar, whereupon we thought them to be tame dogs. There were twenty dogs like mastiffs, with pricked ears and long bushed tails; we found a bone in the pizels of their dogs. Then we went farther and found two sleds made like ours in England. The one was made of fir, spruce, and oaken boards, sawn like inch boards; the other was made all of whalebone, and there hung on the tops of the sleds three heads of beasts which they had killed. We saw here larks, ravens, and partridges.

The 17th we went on shore, and in a little thing made like an oven with stones I found many small trifles, as a small canoe made of wood, a piece of wood made like an image, a bird made of bone, beads having small holes in one end of them to hang about their necks, and other small things. The coast was very barbarous, without wood or grass. The rocks were very fair, like marble, full of veins of divers colours. We found a seal which was killed not long before, being flayed and hid under stones.

Our captain and master searched still for probabilities of the passage, and first found that this place was all islands with great sounds passing between them.

Secondly, the water remained of one colour with the main ocean without altering.

Thirdly, we saw to the west of those isles three or four whales in a scull, which they judged to come from a westerly sea, because to the eastward we saw not any whale.

Also, as we were rowing into a very great sound lying south-west from whence these whales came, upon the sudden there came a violent countercheck of a tide from the south-west against the flood which we came with, not knowing from whence it was maintained.

Fifthly, in sailing 20 leagues within the mouth of this entrance we had sounding in 90 fathoms, fair, grey, oozy sand, and the farther we run into the westwards the deeper was the water, so that hard aboard the shore among these isles we could not have ground in 330 fathoms.

Lastly, it did ebb and flow six or seven fathom up and down, the flood coming from divers parts, so as we could not perceive the chief maintenance thereof.

The 18th and 19th our captain and master determined what was best to do, both for the safe guard of their credits and satisfy of the adventurers, and resolved if the weather brake up to make further search.

The 20th, the wind came directly against us, so they altered their purpose, and reasoned both for proceeding and returning.

The 21st, the wind being north-west, we departed from these islands, and as we coasted the south shore we saw many fair sounds, whereby we were persuaded that it was no firm land but islands.

The 23rd of this month the wind came south-east, very stormy and foul weather. So we were constrained to seek harbour upon the south coast of this entrance, where we fell into a very fair sound, and anchored in 25 fathoms of green, oozy sand, where we went on shore, where we had manifest signs of people, where they had made their fire, and laid stones like a wall. In this place we saw four very fair falcons, and Master Bruton took from one of them his prey, which we judged by the wings and legs to be a snipe, for the head was eaten off.

The 24th, in the afternoon, the wind coming somewhat fair, we departed from this road, purposing by God's grace to return for England.

The 26th we departed from sight of the north land of this entrance, directing our course homewards, until the 10th of the next month.

The 10th September we fell with the Land of Desolation, thinking to go on shore, but we could get never a good harbour. That night we put to sea again thinking to search it the next day; but this night arose a very great storm, and separated our ships so that we lost the sight of the *Moonshine*.

The 13th about noon (having tried all the night before with a goose wing) we set sail, and within two hours after we had sight of the *Moonshine* again. This day we departed from this land.

The 27th of this month we fell with sight of England. This night we had a marvellous storm, and lost the *Moonshine*.

The 30th September we came into Dartmouth, where we found the *Moonshine*, being come in not two hours before.

THE SECOND VOYAGE ATTEMPTED BY MASTER JOHN DAVIS,

With others, for the discovery of the North-West Passage, in Anno 1586.

The 7th day of May I departed from the port of Dartmouth for the discovery of the North-West Passage with a ship of a 120 tons, named the *Mermaid*; a barque of 60 tons, named the *Sunshine*; a barque of 35 tons named the *Moonlight*; and a pinnace of 10 tons named the *North Star*.

And the 15th June I discovered land, in the latitude of 60 degrees, and in longitude from the meridian of London westward 47 degrees, mightily pestered with ice and snow, so that there was no hope of landing; the ice lay in some places 10 leagues, in some 20, and in some 50 leagues off the shore, so that we were constrained to bear into 57 degrees to double the same, and to recover a free sea, which through God's favourable mercy we at length obtained.

The nine-and-twentieth day of June, after many tempestuous storms, we again discovered land in longitude from the meridian of London 58 degrees 30 minutes, and in latitude 64 being east from us, into which course, since it pleased God by contrary winds to force us, I thought it very necessary to bear in with it, and there to set up our pinnace, provided in the *Mermaid* to be our scout for this discovery, and so much the rather, because the year before I had been in the same place and found it very convenient for such a purpose, well stored with float wood, and possessed by a people of tractable conversation; so that the nine-and-twentieth of this month we arrived within the isles which lay before this land, lying north-north-west and south-south-east we know not how far. This land is very high and mountainous, having before it on the west side a mighty company of isles full of fair sounds and harbours. This land was very little troubled with snow, and the sea altogether void of ice.

The ships being within the sounds we sent our boats to search for shallow water, where we might anchor, which in this place is very hard to find; and as the boat went sounding and searching, the people of the country having espied them, came in their canoes towards them with many shouts and cries; but after they had espied in the boat some of our company that were the year before here with us, they presently rowed to the boat and took hold in the oar, and hung about the boat with such comfortable joy as would require a long discourse to be uttered; they came with the boats to our ships, making signs that they knew all those that the year before had been with them. After I perceived their joy and small fear of us, myself with the merchants and others of the company went ashore, bearing with me twenty knives. I had no sooner landed, but they leapt out of their canoes and came running to me and the rest, and embraced us with many signs of hearty welcome. At this present there were eighteen of them, and to each of them I gave a knife; they offered skins to me for reward, but I made signs that it was not sold, but given them of courtesy, and so dismissed them for that time, with signs that they should return again after certain hours.

The next day, with all possible speed, the pinnace was landed upon an isle there to be finished to serve our purpose for the discovery, which isle was so convenient for that purpose, as that we were very well able to defend ourselves against many enemies. During the time that the pinnace was there setting up, the people came continually unto us, sometimes a hundred canoes at a time, sometimes forty, fifty, more and less as occasion served. They brought with them seal skins, stags' skins, white hares, seal fish, salmon peel, small cod, dry caplin, with other fish and birds such as the country did yield.

Myself, still desirous to have a farther search of this place, sent one of the ship boats to one part of the land, and myself went to another part to search for the habitation of this people, with straight commandment that there should be no injury offered to any of the people, neither any one shot.

The boats that went from me found the tents of the people made with seal skins set up upon timber, wherein they found great store of dried caplin, being a little fish no bigger than a pilchard. They found bags of train oil, many little images cut in wood, seal skins in tan tubs with many other such trifles, whereof they diminished nothing.

They also found ten miles within the snowy mountains a plain champion country, with earth and grass, such as our moory and waste grounds of England are. They went up into a river (which in the narrowest place is two leagues broad) about ten leagues, finding it still to continue they knew not how far; but I with my company took another river, which although at the first it offered a large inlet, yet it proved but a deep bay, the end whereof in four hours I attained, and there leaving the boat well manned, went with the rest of my company three or four miles into the country, but found nothing, nor saw anything, save only gripes, ravens, and small birds, as lark and linnet.

The 3rd of July I manned my boat, and went with fifty canoes attending upon me up into another sound, where the people by signs willed me to go, hoping to find their habitation; at length they made signs that I should go into a warm place to sleep, at which place I went on shore, and ascended the top of high hill to see into the country, but perceiving my labour vain, I returned again to my boat, the people still following me and my company very diligent to attend us, and to help us up the rocks, and likewise down; at length I was desirous to have our men leap with them, which was done, but our men did overleap them; from leaping they went to wrestling; we found them strong and nimble, and to have skill in wrestling, for they cast some of our men that were good wrestlers. The 4th of July we launched our pinnace, and had forty of the people to help us, which they did very willingly. At this time our men again wrestled with them, and found them as before, strong and skilful. This 4th of July, the master of the *Mermaid* went to certain islands to store himself with wood, where he found a grave with divers buried in it, only covered with seal skins, having a cross laid over them. The people are of good stature, well in body proportioned, with small, slender hands and feet, with broad visages, and small eyes, wide mouths, the most part unbearded, great lips, and close toothed. Their custom is, as often as they go from us, still at their return, to make a new truce, in this sort: holding his hand up to the sun, with a loud voice crieth "Ylyaoute," and striketh his breast, with like signs being promised safety, he giveth credit. These people are much given to bleed, and therefore stop their noses with deer hair or the hair of an elan. They are idolaters, and have images great store, which they wear about them, and in their boats, which we suppose they worship. They are witches, and have many kinds of enchantments, which they often used, but to small purpose, thanks be to God.

Being among them at shore, the 4th of July, one of them, making a long oration, began to kindle a fire, in this manner: he took a piece of a board, wherein was a hole half through; unto that hole he puts the end of a round stick, like unto a bed staff, wetting the end thereof in train, and in fashion of a turner, with a piece of leather, by his violent motion doth very speedily produce fire; which done, with turfs he made a fire, into which, with many words and strange gestures, he put divers things which we suppose to be a sacrifice. Myself and divers of my company standing by, they were desirous to have me go into the smoke; I willed them likewise to stand in the smoke, in which they by no means would do. I then took one of them, and thrust him into the smoke, and willed one of my company to tread out the fire, and to spurn it into the sea, which was done to show them that we did contemn their sorcery. These people are very simple in all their conversation, but marvellous thievish, especially for iron, which they have in great account. They began through our lenity to show their vile nature; they began to cut our cables; they cut away the *Moonlight's* boat from her stern; they cut our cloth where it lay to air, though we did carefully look unto it, they stole our oars, a calliver, a boat's spear, a sword, with divers other things, whereat the company and masters being grieved, for our better security desired me to dissolve this new friendship, and to leave the company of these thievish miscreants; whereupon there was a calliver shot among them, and immediately upon the same a falcon, which strange noise did sore amaze them, so that with speed they departed; notwithstanding, their simplicity is such, that within ten hours after they came again to us to entreat peace; which, being promised, we again fell into a great league. They brought us seal skins and salmon peel, but, seeing iron, they could in nowise forbear stealing; which, when I perceived it, did but minister unto me an occasion of laughter to see their simplicity, and willed that in no case they should be any more hardly used, but that our own company should be the more vigilant to keep their things, supposing it to be very hard in so short time to make them know their evils. They eat all their meat raw, they live most upon fish, they drink salt water, and eat grass and ice with delight; they are never out of the water, but live in the nature of fishes, but only when dead sleep taketh them, and then under a warm rock, laying his boat upon the land, he lieth down to sleep. Their weapons are all darts, but some of them have bow and arrows and slings. They make nets to take their fish of the fin of a whale; they do all their things very artfully, and it should seem that these simple, thievish islanders have war with those of the main, for many of them are sore wounded, which wounds they received upon the main land, as by signs they gave us to understand. We had among them copper ore, black copper, and red copper; they pronounce their language very hollow, and deep in the throat; these words following we learned from them:—

Kesinyoh, eat some.	Mysacoah, wash it.
Madlycoyte, music.	Lethicksaneg, a seal-skin.
Aginyoh, go, fetch.	Canyglow, kiss me.
Yliaoute, I mean no harm.	Ugnera, my son.
Ponameg, a boat.	Acu, shot.
Conah, leap.	Aba, fallen down.
Maatuke, fish.	Icune, come hither.
Sambah, below.	Awennye, yonder.
Maconmeg, will you have this?	Nugo, no.
Cocah, go to him.	Tucktodo, a fog.
Paaotyck, an oar.	Lechiksah, a skin.
Asanock, a dart.	Maccoah, a dart.
Sawygmeg, a knife.	Sugnacoon, a coat.
Uderah, a nose.	Gounah, come down.
Aoh, iron.	Sasobneg, a bracelet.
Blete, an eye.	Ugnake, a tongue.

Unvicke, give it.	Ataneg, a meal.
Tuckloak, a stag or elan.	Macuah, a beard.
Panygmah, a needle.	Pignagogah, a thread.
Aob, the sea.	Quoysah, give it to me.

The 7th of July, being very desirous to search the habitation of this country, I went myself with our new pinnace into the body of the land, thinking it to be a firm continent, and passing up a very large river a great flaw of wind took me, whereby we were constrained to seek succour for that night, which being had, I landed with the most part of my company, and went to the top of a high mountain, hoping from thence to see into the country; but the mountains were so many and so mighty as that my purpose prevailed not, whereupon I again returned to my pinnace, and willing divers of my company to gather mussels for my supper, whereof in this place there was great store, myself having espied a very strange sight, especially to me, that never before saw the like, which was a mighty whirlwind, taking up the water in very great quantity, furiously mounting it into the air, which whirlwind was not for a puff or blast, but continual for the space of three hours, with very little intermission, which since it was in the course that I should pass, we were constrained that night to take up our lodging under the rocks.

The next morning, the storm being broken up, we went forward in our attempt, and sailed into a mighty great river, directly into the body of the land, and in brief found it to be no firm land, but huge, waste, and desert isles with mighty sounds and inlets passing between sea and sea. Whereupon we returned towards our ships, and landing to stop a flood, we found the burial of these miscreants; we found of their fish in bags, plaices, and caplin dried, of which we took only one bag and departed. The 9th of this month we came to our ships, where we found the people desirous in their fashion of friendship and barter: our mariners complained heavily against the people, and said that my lenity and friendly using of them gave them stomach to mischief, for "they have stolen an anchor from us. They have cut our cable very dangerously, they have cut our boats from our stern, and now, since your departure, with slings they spare us not with stones of half a pound weight. And will you still endure these injuries? It is a shame to bear them." I desired them to be content, and said I doubted not but all should be well. The 10th of this month I went to the shore, the people following me in their canoes; I tolled them on shore, and used them with much courtesy, and then departed aboard, they following me and my company. I gave some of them bracelets, and caused seven or eight of them to come aboard, which they did willingly; and some of them went into the top of our ship, and thus courteously using them I let them depart. The sun was no sooner down but they began to practise their devilish nature, and with slings threw stones very fiercely into the *Moonlight* and struck one of her men, the boatswain, that he overthrew withal: whereat being moved, I changed my courtesy and grew to hatred; myself in my own boat well manned with shot, and the barques boat likewise pursued them, and gave them divers shot, but to small purpose, by reason of their swift rowing; so small content we returned.

The 11th of this month there came five of them to make a new truce; the master of the *Admiral* came to me to show me of their coming, and desired to have them taken and kept as prisoners until we had his anchor again; but when he saw that the chief ring-leader and master of mischief was one of the five, then was vehement to execute his purpose, so it was determined to take him; he came crying "Yliaout," and striking his breast offered a pair of gloves to sell; the master offered him a knife for them: so two of them came to us; the one was not touched, but the other was soon captive among us; then we pointed to him and his fellows for our anchor, which being had we made signs that he should he set at liberty within one hour that he came aboard; the wind came fair, whereupon we weighed and set sail, and so brought the fellow with us. One of his fellows still following our ship close aboard, talked with him, and made a kind of lamentation, we still using him well, with "Yliaout," which was the common course of courtesy. At length this fellow aboard us spoke four or five words unto the other and clapped his two hands upon his face, whereupon the other doing the like, departed, as we supposed, with heavy cheer. We judged the covering of his face with his hands, and bowing of his body down, signified his death. At length he became a pleasant companion among us. I gave him a new suit of frieze after the English fashion, because I saw he could not endure the cold, of which he was very joyful; he trimmed up his darts, and all his fishing tools, and would make oakum, and set his hand to a rope's end upon occasion. He lived with the dry caplin that I took when I was searching in the pinnace, and did eat dry new land fish.

All this while, God be thanked, our people were in very good health, only one young man excepted, who died at sea the 14th of this month, and the 15th, according to the order of the sea, with praise given to God by service, was cast overboard.

The 17th of this month, being in the latitude of 63 degrees 8 minutes, we fell upon a most mighty and strange quantity of ice, in one entire mass, so big as that we knew not the limits thereof, and being withal so very high, in form of a land, with bays and capes, and like high cliff land as that we supposed it to be land, and therefore sent our pinnace off to discover it; but at her return we were certainly informed that it was only ice, which bred great admiration to us all, considering the huge quantity thereof incredible to be reported in truth as it was, and therefore I omit to speak any further thereof. This only, I think that the like before was never seen, and in this place we had very stickle and strong currents.

We coasted this mighty mass of ice until the 30th of July, finding it a mighty bar to our purpose: the air in this time was so contagious, and the sea so pestered with ice, as that all hope was banished of proceeding; for the 24th of July all our shrouds, ropes, and sails were so frozen, and encompassed with ice, only by a gross fog, as seemed to be more than strange, since the last year I found this sea free and navigable, without impediments.

Our men through this extremity began to grow sick and feeble, and withal hopeless of good success; whereupon, very orderly, with good discretion they entreated me to regard the state of this business, and withal advised me that in conscience I ought to regard the safety of mine own life with the preservation of theirs, and that I should not, through my overboldness, leave their widows and fatherless children to give me bitter curses. This matter in conscience did greatly move me to regard their estates, yet considering the excellency of the business, if it might be obtained, the great hope of certainty by the last year's discovery, and that there was yet a third way not put in practice, I thought it would grow to my disgrace if this action by my negligence should grow into discredit: whereupon seeking help from God, the fountain of all mercies, it pleased His Divine Majesty to move my heart to prosecute that which I hope shall be to His glory, and to the contentation of every Christian mind. Whereupon, falling into consideration that the *Mermaid*, albeit a very strong and sufficient ship, yet by reason of her burden not so convenient and nimble as a smaller barque, especially in such desperate hazards; further, having in account how great charge to the adventurers, being at 100 livres the month, and that in doubtful service, all the premises considered, with divers other things, I determined to furnish the *Moonlight* with revictualing and sufficient men, and to proceed in this action as God should direct me; whereupon I altered our course from the ice, and bore east-south-east to the cover of the next shore, where this thing might be performed; so with favourable wind it pleased God that the 1st of August we discovered the land in latitude 66 degrees 33 minutes, and in longitude from the meridian of London 70 degrees, void of trouble, without snow or ice.

The 2nd of August we harboured ourselves in a very excellent good road, where with all speed we graved the *Moonlight*, and revictualled her; we searched this country with our pinnace while the barque was trimming, which William Eston did: he found all this land to be only islands, with a sea on the east, a sea on the west, and a sea on the north. In this place we found it very hot, and we were very much troubled with a fly which is called mosquito, for they did sting grievously. The people of this place at our first coming in caught a seal, and, with bladders fast tied to him sent him in to us with the flood, so as he came right with our ships, which we took as a friendly present from them.

The 5th of August I went with the two masters and others to the top of a hill, and by the way William Eston espied three canoes lying under a rock, and went unto them: there were in them skins, darts, with divers superstitious toys, whereof we diminished no thing, but left upon every boat a silk point, a bullet of lead, and a pin. The next day, being the 6th of August, the people came unto us without fear, and did barter with us for skins, as the other people did: they differ not from the other, neither in their canoes nor apparel, yet is their pronunciation more plain than the others, and nothing hollow in the throat. Our miscreant aboard of us kept himself close, and made show that he would fain have another companion. Thus being provided, I departed from this land the 12th of August at six of the clock in the morning, where I left the *Mermaid* at anchor; the 14th sailing west about 50 leagues we discovered land, being in latitude 66 degrees 19 minutes: this land is 70 leagues from the other from whence we came. This 14th day, from nine o'clock at night till three o'clock in the morning, we anchored by an island of ice 12 leagues off the shore, being moored to the ice.

The 15th day, at three o'clock in the morning, we departed from this land to the south, and the 18th of August we discovered land north-west from us in the morning, being a very fair promontory, in latitude 65 degrees, having no land on the south. Here we had great hope of a through passage.

This day, at three o'clock in the afternoon, we again discovered land south-west and by south from us, where at night we were becalmed. The 19th of this month at noon, by observation, we were in 64 degrees 20 minutes. From the 18th day at noon until the 19th at noon, by precise ordinary care, we had sailed fifteen leagues south and by west, yet by art and more exact observation we found our course to be south-west, so that we plainly perceived a great current striking to the west.

This land is nothing in sight but isles, which increaseth our hope. This 19th of August, at six o'clock in the afternoon, it began to snow, and so continued all night, with foul weather and much wind, so that we were constrained to lie at hull all night, five leagues off the shore: in the morning, being the 20th of August, the fog and storm breaking up, we bore in with the land, and at nine o'clock in the morning we anchored in a very fair and safe road and locket for all weathers. At ten o'clock I went on shore to the top of a very high hill, where I perceived that this land was islands; at four o'clock in the afternoon we weighed anchor, having a fair north-north-east wind, with very fair weather; at six o'clock we were clear without the land, and so shaped our course to the south, to discover the coast whereby the passage may be through God's mercy found.

We coasted this land till the 28th day of August, finding it still to continue towards the south, from the latitude of 67 to 57 degrees; we found marvellous great store of birds, gulls and mews, incredible to be reported, whereupon being calm weather we lay one glass upon the lee to prove for fish, in which space we caught one hundred of cod, although we were but badly provided for fishing, not being our purpose. This 28th, having great distrust of the weather, we arrived in a very fair harbour in the latitude of 56 degrees, and sailed ten leagues in the same, being two leagues broad, with very fair woods on both sides; in this place we continued until the 1st of September, in which time we had two very great storms. I landed, and went six miles by guess into the country, and found that the woods were fir, pine-apple, alder, yew, withy, and birch; here we saw a black bear; this place yieldeth great store of birds, as pheasant, partridge, Barbary hens, or the like, wild geese, ducks, blackbirds, jays, thrushes, with other kinds of small birds. Of the partridge and pheasant we killed great store with bow and arrows in this place; at the harbour-mouth we found great store of cod.

The 1st of September at ten o'clock we set sail, and coasted the shore with very fair weather. The third day being calm, at noon we struck sail, and let fall a cadge anchor to prove whether we could take any fish, being in latitude 54 degrees 30 minutes, in which place we found great abundance of cod, so that the hook was no sooner overboard but presently a fish was taken. It was the largest and best refet fish that ever I saw, and divers fishermen that were with me said that they never saw a more suaule, or better skull of fish in their lives, yet had they seen great abundance.

The 4th of September, at 5 o'clock in the afternoon, we anchored in a very good road among great store of isles, the country low land, pleasant, and very full of fair woods. To the north of this place eight leagues we had a perfect hope of the passage, finding a mighty great sea passing between two lands west. The south land to our judgment being nothing but isles, we greatly desired to go into this sea, but the wind was directly against us. We anchored in four fathom fine sand.

In this place is fowl and fish mighty store.

The 6th of September, having a fair north-north-west wind, having trimmed our barque, we purposed to depart, and sent five of our sailors, young men, ashore to an island to fetch certain fish which we purposed to weather, and therefore left it all night covered upon the isle; the brutish people of this country lay secretly lurking in the wood, and upon the sudden assaulted our men, which when we perceived, we presently let slip our cables upon the halse, and under our foresail bore into the shore, and with all expedition discharged a double musket upon them twice, at the noise whereof they fled; notwithstanding, to our very great grief, two of our men were slain with their arrows, and two grievously wounded, of whom, at this present, we stand in very great doubt; only one escaped by swimming, with an arrow shot through his arm. These wicked miscreants never offered parley or speech, but presently executed their cursed fury. This present evening it pleased God farther to increase our sorrows with a mighty tempestuous storm, the wind being north-north-east, which lasted unto the 10th of this month very extreme. We unrigged our ship, and purposed to cut-down our masts; the cable of our shut anchor broke, so that we only expected to be driven on shore amongst these cannibals for their prey. Yet in this deep distress the mighty mercy of God, when hope was past, gave us succour, and sent us a fair lee, so as we recovered our anchor again, and new-moored our ship; where we saw that God manifestly delivered us, for the strains of one of our cables were broken; we only rode by an old junk. Thus being freshly moored, a new storm arose, the wind being west-north-west, very forcible, which lasted unto the 10th day at night.

The 11th day, with a fair west-north-west wind, we departed with trust in God's mercy, shaping our course for England, and arrived in the West Country in the beginning of October.

* * * * *

Master Davis being arrived, wrote his letter to Master William Sanderson of London, concerning his voyage, as followeth.

> Sir,—The *Sunshine* came into Dartmouth the 4th of this month: she hath been at Iceland, and from thence to Greenland, and so to Estotiland, from thence to Desolation, and to our merchants, where she made trade with the people, staying in the country twenty days. They have brought home 500 seal-skins, and 140 half skins and pieces of skins. I stand in great doubt of the pinnace; God be merciful unto the poor men and preserve them if it be His blessed will.
>
> I have now full experience of much of the north-west part of the world, and have brought the passage to that certainty, as that I am sure it must be in one of four places, or else not at all. And further, I can assure you upon the peril of my life, that this voyage may be performed without further charge, nay, with certain profit to the adventurers, if I may have but your favour in the action. Surely it shall cost me all my hope of welfare and my portion of Sandridge, but I will, by God's mercy, see an end of these businesses. I hope

I shall find favour with you to see your card. I pray God it be so true as the card shall be which I will bring to you, and I hope in God that your skill in navigation shall be gainful unto you, although at the first it hath not proved so. And thus with my most humble commendations I commit you to God, desiring no longer to live than I shall be yours most faithfully to command. From this 14th of October, 1586.

Yours with my heart, body and life to command,

JOHN DAVIS.

* * * * *

The relation of the course which the "Sunshine," a barque of fifty tons, and the "North Star," a small pinnace, being two vessels of the fleet of Master John Davis, held after he had sent them from him to discover the passage between Greenland and Iceland. Written by Henry Morgan, servant to Master William Sanderson of London.

The 7th day of May, 1586, we departed out of Dartmouth Haven four sails, to wit, the *Mermaid*, the *Sunshine*, the *Moonshine*, and the *North Star*. In the *Sunshine* were sixteen men, whose names were these: Richard Pope, master; Mark Carter, master's mate; Henry Morgan, purser; George Draward, John Mandie, Hugh Broken, Philip Jane, Hugh Hempson, Richard Borden, John Filpe, Andrew Madocke, William Wolcome, Robert Wagge, carpenter, John Bruskome, William Ashe, Simon Ellis.

Our course was west-north-west the 7th and 8th days; and the ninth day in the morning we were on head of the Tarrose of Scilly. Thus coasting along the south part of Ireland, the 11th day we were on the head of the Dorses, and our course was south-south-west until six of the clock the 12th day. The 13th day our course was north-west. We remained in the company of the *Mermaid* and the *Moonshine* until we came to the latitude of 60 degrees, and there it seemed best to our general, Master Davis, to divide his fleet, himself sailing to the north-west, and to direct the *Sunshine*, wherein I was, and the pinnace called the *North Star*, to seek a passage northward between Greenland and Iceland to the latitude of 80 degrees, if land did not let us. So the 7th day of June we departed from them, and the 9th of the same we came to a firm land of ice, which we coasted along the 9th, the 10th, and the 11th days of June; and the 11th day at six of the clock at night we saw land, which was very high, which afterwards we knew to be Iceland, and the 12th day we harboured there, and found many people; the land lieth east and by north in 66 degrees.

Their commodities were green fish and Iceland lings and stock fish, and a fish which is called catfish, of all which they had great store. They had also kine, sheep, and horses, and hay for their cattle and for their horses. We saw also of their dogs. Their dwelling-houses were made on both sides with stones, and wood laid across over them, which was covered over with turfs of earth, and they are flat on the tops, and many of these stood hard by the shore. Their boats were made with wood, and iron all along the keel like our English boats; and they had nails for to nail them withal, and fish-hooks, and other things for to catch fish as we have here in England. They had also brazen kettles, and girdles and purses made of leather, and knops on them of copper, and hatchets, and other small tools as necessary as we have. They dry their fish in the sun; and when they are dry they pack them up in the top of their houses. If we would go thither to fishing more than we do, we should make it a very good voyage, for we got a hundred green fishes in one morning. We found here two Englishmen with a ship, which came out of England about Easter Day of this present year, 1586; and one of them came aboard of us and brought us two lambs. The Englishman's name was Master John Royden, of Ipswich, merchant; he was bound for London with his ship. And this is the sum of that which I observed in Iceland. We departed from Iceland the 16th day of June, in the morning, and our course was north-west; and saw on the coast two small barques going to a harbour; we went not to them, but saw them afar off. Thus we continued our course unto the end of this month.

The 3rd day of July we were in between two firm lands of ice, and passed in between them all that day until it was night, and then the master turned back again, and so away we went towards Greenland. And the 7th day of July we did see Greenland, and it was very high, and it looked very blue; but we could not come to harbour in the land because we were hindered by a firm land, as it were, of ice, which was along the shore's side; but we were within three leagues of the land, coasting the same divers days together. The 17th day of July we saw the place which our captain, Master John Davis, the year before had named the Land of Desolation, where we could not go on shore for ice. The 18th day we were likewise troubled with ice, and went in amongst it at three of the clock in the morning. After we had cleared ourselves thereof we ranged all along the coast of Desolation until the end of the aforesaid month.

The 3rd day of August we came in sight of Gilbert's Sound in the latitude of 64 degrees 15 minutes, which was the place where we were appointed to meet our general and the rest of our fleet. Here we came to a harbour at six of the clock at night.

The 4th day, in the morning, the master went on shore with ten of his men, and they brought us four of the people rowing in their boats, aboard of the ship. And in the afternoon I went on shore with six of our men, and there came to us seven of them when we were on land. We found on shore three dead people, and two of them had their staves lying by them, and their old skins wrapped about them, and the other had nothing lying by, wherefore we thought it was a woman. We also saw their houses, near the seaside, which were made with pieces of wood on both sides, and crossed over with poles and then covered over with earth. We found foxes running upon the hills. As for the place, it is broken land all the way that we went, and full of broken islands. The 21st of August the master sent the boat on shore for wood, with six of his men, and there were one-and-thirty of the people of the country, which went on shore to them, and they went about to kill them as we thought, for they shot their darts towards them, and we that were aboard the ship did see them go on shore to our men, whereupon the master sent the pinnace after them; and when they saw the pinnace coming towards them they turned back, and the master of the pinnace did shoot off a culliver to them the same time, but hurt none of them, for his meaning was only to put them in fear. Divers times they did wave us on shore to play with them at the football, and some of our company went on shore to play with them, and our men did cast them down as soon as they did come to strike the ball. And thus much of that which we did see and do in that harbour where we arrived first.

The 23rd day we departed from the merchants where we had been first, and our course from thence was south and by west, and the wind was north-east, and we ran that day and night about five or six leagues until we came to another harbour.

The 24th, about eleven of the clock in the forenoon, we entered into the aforesaid new harbour, and as we came in we did see dogs running upon the islands. When we were come in, there came to us four of the people which were with us before in the other harbour; and where we rowed we had sandy ground. We saw no wood growing, but found small pieces of wood upon the islands, and some small pieces of sweet wood among the same. We found great harts' horns, but could see none of the stags where we went, but we found their footings. As for the bones which we received of the savages, I cannot tell of what beasts they be. The stones that we found in the country were black, and some white; as I think, they be of no value; nevertheless I have brought examples of them to you.

The 30th of August we departed from this harbour towards England, and the wind took us contrary, so that we were fain to go to another harbour the same day at eleven of the clock. And there came to us thirty-nine of the people and brought us thirteen seal-skins, and after we received these skins of them the master sent the carpenter to change one of our boats which we had bought of them before; and they would have taken the boat from him perforce, and when they saw they could not take it from us they shot with their darts at us, and struck one of our men with one of their darts, and John Filpe shot one of them in the breast with an arrow. And they came to us again, and four of our men went into the ship boat, and they shot with their darts at our men; but our men took one of their people in his boat, into the ship boat, and he hurt one of them with his knife, but we killed three of them in their boats, two of them were hurt with arrows in the breast, and he that was aboard our boat was shot with an arrow, and hurt with a sword, and beaten with staves, whom our men cast overboard; but the people caught him and carried him on shore upon their boats, and the other two also, and so departed from us. And three of them went on shore hard by us where they had their dogs, and those three came away from their dogs, and presently one of their dogs came swimming towards us hard aboard the ship, whereupon our master caused the gunner to shoot off one of the great pieces—towards the people, and so the dog turned back to land, and within an hour after there came of the people hard aboard the ship, but they would not come to us as they did before.

The 31st of August we departed from Gilbert's Sound for England, and when we came out of the harbour there came after us seventeen of the people looking which way we went.

The 2nd of September we lost sight of the land at twelve of the clock at noon.

The 3rd day at night we lost sight of the *North Star*, our pinnace, in a very great storm, and lay a-hull tarrying for them the 4th day, but could hear no more of them. Thus we shaped our course the 5th day south-south-east, and sailing unto the 27th of the said month, we came in sight of Cape Clear in Ireland.

The 30th day we entered into our own Channel.

The 2nd of October we had sight of the Isle of Wight.

The 3rd we coasted all along the shore, and the 4th and 5th.

The 6th of the said month of October we came into the River of Thames as high as Ratcliffe in safety, God be thanked!

THE THIRD VOYAGE NORTH-WESTWARD, MADE BY JOHN DAVIS,

Gentleman, as chief captain and pilot general for the discovery of a passage to the Isles of the Molucca, or the coast of China, in the year 1587. *Written by John Janes, servant to the aforesaid Master William Sanderson.*

May.—The 19th of this present month, about midnight, we weighed our anchors, set sail and departed from Dartmouth with two barques and a clincher, the one named the *Elizabeth*, of Dartmouth, the other the *Sunshine*, of London, and the clincher called the *Ellin*, of London; thus, in God's name, we set forwards with wind at north-east, a good fresh gale. About three hours after our departure, the night being somewhat thick with darkness, we had lost the pinnace. The captain, imagining that the men had run away with her, willed the master of the *Sunshine* to stand to seawards and see if we could descry them, we bearing in with the shore for Plymouth. At length we descried her, bore with her, and demanded what the cause was; they answered that the tiller of their helm was burst, so shaping our course west-south-west, we went forward, hoping that a hard beginning would make a good ending; yet some of us were doubtful of it, failing in reckoning that she was a clincher; nevertheless, we put our trust in God.

The 21st we met with the *Red Lion* of London, which came from the coast of Spain, which was afraid that we had been men-of-war; but we hailed them, and after a little conference we desired the master to carry our letters for London, directed to my uncle Sanderson, who promised us safe delivery. And after we had heaved them a lead and a line, whereunto we had made fast our letters, before they could get them into the ship they fell into the sea, and so all our labour and theirs also was lost; notwithstanding, they promised to certify our departure at London, and so we departed, and the same day we had sight of Scilly. The 22nd the wind was at north-east by east, with fair weather, and so the 23rd and 24th the like. The 25th we laid our ships on the lee for the *Sunshine*, who was a-rummaging for a leak; they had 500 strokes at the pump in a watch, with the wind at north-west.

The 26th and 27th we had fair weather, but this 27th the pinnace's foremast was blown overboard. The 28th the *Elizabeth* towed the pinnace, which was so much bragged of by the owner's report before we came out of England, but at sea she was like a cart drawn with oxen. Sometimes we towed her, because she could not sail for scant wind.

The 31st day our captain asked if the pinnace were staunch. Peerson answered that she was as sound and staunch as a cup. This made us something glad when we saw she would brook the sea, and was not leaky.

June.—The first six days we had fair weather; after that for five days we had fog and rain, the wind being south.

The 12th we had clear weather. The mariners in the *Sunshine* and the master could not agree; the mariners would go on their voyage a-fishing, because the year began to waste; the master would not depart till he had the company of the *Elizabeth*, whereupon the master told our captain that he was afraid his men would shape some contrary course while he was asleep, and so he should lose us. At length, after much talk and many threatenings, they were content to bring us to the land which we looked for daily.

The 13th we had fog and rain.

The 14th day we discovered land at five of the clock in the morning, being very great and high mountains, the tops of the hills being covered with snow. Here the wind was variable, sometimes north-east, east-north-east, and east by north; but we imagined ourselves to be 16 or 17 leagues off from the shore.

The 15th we had reasonably clear weather.

The 16th we came to an anchor about four or five of the clock in the afternoon. The people came presently to us, after the old manner, with crying "Il y a oute," and showed us seal-skins.

The 17th we began to set up the pinnace that Peerson framed at Dartmouth, with the boards which he brought from London.

The 18th, Peerson and the carpenters of the ships began to set on the planks.

The 19th, as we went about an island, were found black pumice stones, and salt kerned on the rocks, very white and glistering. This day, also, the master of the *Sunshine* took one of the people, a very strong, lusty young fellow.

The 20th, about two of the clock in the morning, the savages came to the island where our pinnace was built ready to be launched, and tore the two upper strakes and carried them away, only for the love of the iron in the boards. While they were about this practice, we manned the *Elizabeth's* boat to go ashore to them. Our men, being either afraid or amazed, were so long before they came to shore, that our captain willed them to stay, and made the gunner give fire to a saker, and laid the piece level with the boat, which the savages had turned on the one side because we could not hurt them with our arrows, and made the boat their bulwark against the arrows which we shot at them. Our gunner, having made all things ready, gave fire to the piece, and fearing to hurt any of the people, and regarding the owner's profit, thought belike he would save a saker's shot, doubting we should have occasion to fight with men-of-war, and so shot off the saker without a bullet, we looking still when the savages that were hurt should run away without legs; at length we could perceive never a man hurt, but all having their legs, could carry away their bodies. We had no sooner shot off the piece but the master of the *Sunshine* manned his boat, and came rowing towards the island, the very sight of whom made each of them take that he had gotten, and fly away as fast as they could to another island about two miles off, where they took the nails out of the timber, and left the wood on the isle. When we came on shore, and saw how they had spoiled the boat, after much debating of the matter, we agreed that the *Elizabeth* should have her to fish withal; whereupon she was presently carried aboard and stowed. Now after this trouble, being resolved to depart with the first wind, there fell out another matter worse than all the rest, and that was in this manner: John Churchyard, one whom our captain had appointed as pilot in the pinnace, came to our captain and Master Bruton, and told them that the good ship which we must all hazard our lives in had three hundred strokes at one time as she rode in the harbour. This disquieted us all greatly, and many doubted to go in her. At length our captain, by whom we were all to be governed, determined rather to end his life with credit than to return with infamy and disgrace; and so, being all agreed, we purposed to live and die together, and committed ourselves to the ship.

Now the 21st, having brought all our things aboard, about eleven or twelve of the clock at night we set sail and departed from those isles, which lie in 64 degrees of latitude, our ships being now all at sea, and we shaping our course to go coasting the land to the northwards, upon the eastern shore, which we called the shore of our merchants, because there we met with people which traffic with us; but here we were not without doubt of our ship.

The 22nd and 23rd we had close fog and rain.

The 24th, being in 67 degrees and 40 minutes, we had great store of whales, and a kind of sea-birds which the mariners call cortinous. This day, about six of the clock at night, we espied two of the country people at sea, thinking at the first they had been two great seals, until we saw their oars, glistering with the sun. They came rowing towards us as fast as they could, and when they came within hearing they held up their oars and cried "Il y a oute," making many signs, and at last they came to us, giving us birds for bracelets, and of them I had a dart with a bone in it, or a piece of unicorn's horn, as I did judge. This dart he made store of, but when he saw a knife he let it go, being more desirous of the knife than of his dart. These people continued rowing after our ship the space of three hours.

The 25th, in the morning, at seven of the clock, we descried thirty savages rowing after us, being by judgment ten leagues off from the shore. They brought us salmon peels, birds, and caplin, and we gave them pins, needles, bracelets, nails, knives, bells, looking-glasses, and other small trifles; and for a knife, a nail, or a bracelet, which they call ponigmah, they would sell their boat, coats, or anything they had, although they were far from the shore. We had but few skins of them, about twenty; but they made signs to us that if we would go to the shore, we should have more store of chicsanege. They stayed with us till eleven of the clock, at which time we went to prayer, and they departed from us.

The 26th was cloudy, the wind being at south.

The 27th fair, with the same wind.

The 28th and 29th were foggy, with clouds.

The 30th day we took the height, and found ourselves in 72 degrees and 12 minutes of latitude, both at noon and at night, the sun being five degrees above the horizon. At midnight the compass set to the variation of 28 degrees to the westward. Now having coasted the land which we called London Coast from the 21st of this present till the 30th, the sea open all to the westwards and northwards, the land on starboard side east from us, the wind shifted to the north, whereupon we left that shore, naming the same Hope Sanderson, and shaped our course west, and ran forty leagues and better without the sight of any land.

July.—The 2nd we fell in with a mighty bank of ice west from us, lying north and south, which bank we would gladly have doubled out to the northwards, but the wind would not suffer us, so that we were fain to coast it to the southwards, hoping to double it out that we might have run so far west till we had found land, or else to have been thoroughly resolved of our pretended purpose.

The 3rd we fell in with the ice again, and putting off from it we sought to the northwards, but the wind crossed us.

The 4th was foggy, so was the 5th; also with much wind at north.

The 6th being very clear, we put our barque with oars through a gap in the ice, seeing the sea free on the west side, as we thought, which falling out otherwise, caused us to return after we had stayed there between the ice.

The 7th and the 8th, about midnight, by God's help we recovered the open sea, the weather being fair and calm; and so was the 9th.

The 10th we coasted the ice.

The 11th was foggy, but calm.

The 12th we coasted again the ice, having the wind at west-north-west. The 13th, bearing off from the ice, we determined to go with the shore, and come to an anchor, and to stay five or six days for the dissolving of the ice, hoping that the sea from continually beating it, and the sun with the extreme force of heat, which it had always shining upon it, would make a quick despatch, that we might have a further search upon the western shore. Now when we were come to the eastern coast, the water something deep, and some of our company fearful withal, we durst not come to an anchor, but bore off into sea again. The poor people, seeing us go away again, came rowing after us into the sea, the waves being somewhat lofty. We trucked with them for a few skins and darts, and gave them beads, nails, needles, and cards, they pointing to the shore as though they would show us great friendship; but we, little regarding their courtesy, gave them the gentle farewell, and so departed.

The 14th we had the wind at south. The 15th there was some fault either in the barque or the set of some current, for we were driven six points out of our course. The 16th we fell in with the bank of ice, west from us. The 17th and 18th were foggy. The 19th, at one o'clock afternoon, we had sight of the land which we called Mount Raleigh, and at twelve of the clock at night we were athwart the straits which we discovered the first year. The 20th we traversed in the mouth of the strait, the wind being at west with fair and clear weather. The 21st and 22nd we coasted the northern coast of the straits. The 23rd, having sailed 60 leagues north-west into the straits at two o'clock afternoon, we anchored among many isles in the bottom of the gulf, naming the same the Earl of Cumberland's Isles, where, riding at anchor, a whale passed by our ship and went west in among the isles. Here the compass set at 30 degrees westward variation. The 24th we departed, shaping our course south-east to recover the sea. The 25th we were becalmed in the bottom of the gulf, the air being extremely hot. Master Bruton and some of the mariners went on shore to course dogs, where they found many graves, and trains spilt on the ground, the dogs being so fat that they were scant able to run.

The 26th we had a pretty storm, the wind being at south-east. The 27th and 28th were fair. The 29th we were clear out of the straits, having coasted the south shore, and this day at noon we were in 64 degrees of latitude. The 30th in the afternoon we coasted a bank of ice which lay on the shore, and passed by a great bank or inlet which lay between 63 and 62 degrees of latitude, which we called Lumley's Inlet. We had oftentimes, as we sailed along the coast, great roots, the water as it were whirling and overfalling, as if it were the fall of some great water through a bridge. The 31st as we sailed by a headland, which we named Warwick's Forehand, we fell into one of those overfalls with a fresh gale of wind, and bearing all our sails, we looking upon an island of ice between us and the shore, had thought that our barque did make no way, which caused us to take marks on the shore. At length we perceived ourselves to go very fast, and the island of ice which we saw before was carried very forcibly with the set of the current faster than our ship went. This day and night we passed by a very great gulf, the water whirling and roaring as it were the meeting of tides.

August.—The 1st, having coasted a bank of ice which was driven out at the mouth of this gulf, we fell in with the southernmost cape of the gulf, which we named Chidlie's Cape, which lay in 6 degrees and 10 minutes of latitude. The 2nd and 3rd were calm and foggy, so were the 4th, 5th, and 6th. The 7th was fair and calm, so was the 8th, with a little gale in the morning. The 9th was fair, and we had a little gale at night. The 10th we had a frisking gale at west-north-west; the 11th fair. The 12th we saw five deer on the top of an island, called by us Darcie's Island. And we hoisted out our boat, and went ashore to them, thinking to have killed some of them. But when we came on shore and had coursed them twice about the island they took the sea, and swain towards islands distant from that three leagues. When we perceived that they had taken the sea, we gave them over, because our boat was so small that it could not carry us and row after them, they swam so fast; but one of them was as big as a good pretty cow, and very fat; their feet as big as ox-feet. Here upon this island I killed with my piece a grey hare.

The 13th in the morning we saw three or four white bears, but durst not go on shore unto them for lack of a good boat. This day we struck a rock seeking for a harbour, and received a leak, and this day we were in 54 degrees of latitude. The 14th we stopped our leak in a storm not very outrageous at noon.

The 15th, being almost in 51 degrees of latitude, and not finding our ships, nor (according to their promise) being any mark, token, or beacon, which we willed to set up, and they protested to do so upon every headland, sea, island, or cape, within 20 leagues every way off from their fishing place, which our captain appointed to be between 54 and 55 degrees—this 15th, I say, we shaped our course homeward for England, having in our ship but little wood, and half a hogshead of fresh water. Our men were very willing to depart, and no man more forward than Peerson, for he feared to be put out of his office of stewardship; he was so insatiate that the allowance of two men was scant sufficient to fill his greedy appetite; but because every man was so willing to depart, and considering our want, I doubted the matter very much, fearing that the seething of our men's victuals in salt water would breed diseases, and being but few (yet too many for the room, if any should be sick), and likely that all the rest might be infected therewith, we consented to return for our own country, and so we had the 16th there with the wind at south-west.

The 17th we met a ship at sea, and as far as we could judge it was a Biscayan; we thought she went a-fishing for whales, for in 52 degrees or thereabout we saw very many.

The 18th was fair with a good gale at west.

The 19th fair also, but with much wind at west and by south.

And thus, after much variable weather and change of winds, we arrived the 15th of September in Dartmouth, Anno 1587, giving thanks to God for our safe arrival.

* * * * *

A letter of the said Master John Davis, written to Master Sanderson of London, concerning his fore-written voyage.

> GOOD MASTER SANDERSON,—With God's great mercy I have made my safe return in health with all my company, and have sailed 60 leagues farther than my determination at my departure. I have been in 73 degrees, finding the sea all open, and 40 leagues between laud and land; the passage is most certain, the execution most easy, as at my coming you shall fully know. Yesterday, the 15th of September, I landed all weary, therefore I pray you pardon my shortness.

Sandridge, this 16th of September, Anno 1587.

> Yours equal as mine own, which
> by trial you shall best know,
> JOHN DAVIS.

www.ingramcontent.com/pod-product-compliance
Ingram Content Group UK Ltd.
Pitfield, Milton Keynes, MK11 3LW, UK
UKHW042150281224
453045UK00004B/295